KNOWLEDGE-BASED SERVICES, INTERNAT
AND REGIONAL DEVELOPME

Knowledge-Based Services, Internationalization and Regional Development

Edited by

JAMES W. HARRINGTON
University of Washington, USA

PETER W. DANIELS
University of Birmingham, UK

Routledge
Taylor & Francis Group

LONDON AND NEW YORK

First published 2006 by Ashgate Publishing

2 Park Square, Milton Park, Abingdon, Oxon OX14 4RN
711 Third Avenue, New York, NY 10017, USA

Routledge is an imprint of the Taylor & Francis Group, an informa business

First issued in paperback 2017

Copyright © James W. Harrington and Peter W. Daniels 2006

James W. Harrington and Peter W. Daniels have asserted their right under the Copyright, Designs and Patents Act, 1988, to be identified as the editors of this work.

All rights reserved. No part of this book may be reprinted or reproduced or utilised in any form or by any electronic, mechanical, or other means, now known or hereafter invented, including photocopying and recording, or in any information storage or retrieval system, without permission in writing from the publishers.

Notice:
Product or corporate names may be trademarks or registered trademarks, and are used only for identification and explanation without intent to infringe.

British Library Cataloguing in Publication Data
Knowledge-based services, internationalization and regional
    development. - (The dynamics of economic space)
    1. Information services industry - Congresses 2. Economic
    geography - Congresses 3. Regional economics - Congresses
    4. Space in economics - Congresses 5. Information
    technology - Congresses 6. Information networks -
    Congresses
    I. Harrington, J. W. (James W.), 1957- II. Daniels, P. W.
    338.4'7004

Library of Congress Control Number: 2006932216

ISBN 978-0-7546-4897-0 (hbk)
ISBN 978-1-138-27557-7 (pbk)

Contents

List of Figures

List of Tables

List of Contributors

Jonathan V. Beaverstock, Professor of Economic Geography, Loughborough University, Loughborough, Leicestershire, LE11 3TU. E-mail: j.v.beaverstock@lboro.ac.uk. His research focuses on: globalization and world cities; the organizational strategies of transnational professional service firms in international financial centres; and international human resource management.

John R. Bryson, Professor of Enterprise and Economic Geography, The School of Geography, Earth and Environmental Sciences, The University of Birmingham, Edgbaston, Birmingham, B15 2TT. E-mail: J.R.Bryson@bham.ac.uk. Professor Bryson's primary research focus is the spatial and organizational dynamics of knowledge-intensive firms, especially business and professional services and their clients.

José Antonio Camacho, Titular Professor, Department of Applied Economics, Campus Universitario de Cartuja s/n, E-18071, Granada (Spain). E-mail: jcamacho@ugr.es. Professor Camacho's research focus is the application of input-output to services. In particular, he studies the tertiarization process through input-output tables.

Peter W. Daniels, Professor of Geography, University of Birmingham, School of Geography, Earth and Environmental Sciences, Edgbaston, Birmingham B15 2TT, UK. E-mail: p.w.daniels@bham.ac.uk. Professor Daniels has undertaken research and published several articles and books on the journey to work impacts of office decentralization, on the location and development of office activities, and on the role of service industries, especially producer services, as key drivers of metropolitan and regional restructuring at the national and international scale.

Vincenzo Demetrio, Post-doctoral Fellow in Economic Geography, University of Turin, Piazza Arbarello 8, 10122, Torino, Italy. E-mail: demetrio@econ.unito.it. Dr. Demetrio's main research field is post-industrial transition in North Italy and local development policies.

H. Peter Dörrenbächer, Prof. Dr, Universität des Saarlandes, Department of Geography, Postfach 151150, 66041 Saarbrücken, Germany. E-mail: p.doerren@mx.uni-saarland.de. Professor Dörrenbächer's research focus is Economic and Social Geography, Cultural Geography, Geography of the EU, Canada, border regions and transboundary relations with a focus on Western Europe.

James R. Faulconbridge, Lecturer in Human Geography, Department of Geography, Lancaster University, Lancaster, LA1 4YW, UK. Email: j.faulconbridge@lancaster. ac.uk. Dr. Faulconbridge's research focus is professional service firms and their globalization. Previous and ongoing research examines a range of sectors (advertising, executive search, law and financial services) and the ways relational organizational forms allow the delivery of integrated global services and the production and circulation of professional knowledges.

Paolo Giaccaria, Lecturer in Political and Economic Geography, University of Turin, Piazza Arbarello 8, 10122, Torino, Italy. E-mail: paolo.giaccaria@unito.it. Dr Giaccaria has written on local development. His main research interests focus currently on the epistemological foundation of Economic Geography and on the political geography of the Mediterranean scale.

Sarah J.E. Hall, Lecturer in Human Geography. Department of Geography, Loughborough University, Loughborough, Leicestershire, LE11 3TU. E-mail: S.J.E.Hall@lboro.ac.uk. Dr Hall studies the relationship between business theory and practice. Previous research has explored this dynamic in relation to London's investment banking industry. Ongoing research projects develop this work in relation to executive education and the changing geographies of the headhunting industry.

James W. Harrington, Professor of Geography, Box 353550, University of Washington, Seattle WA 98195, United States. E-mail: jwh@u.washington.edu. Professor Harrington has written on regional economic development, primarily in North America. His current research focuses on region-specific institutions for leadership and labor force development.

Hyungjoo Kim, Associate Research Fellow, Science and Technology Policy Institute, Specialty Construction Center 26F 395-70, Shindaebang-dong, Dongjak-gu, Seoul, Korea 156-714. E-mail: hjkim@stepi.re.kr. Dr. Kim's research interests lie in innovation and economic development, with particular focus on human capital and higher education.

Jari Kolehmainen, Researcher, Research Unit for Urban and Regional Development Studies; 33014 University of Tampere; Finland. E-mail: jari.kolehmainen@uta.fi. Kolehmainen's research focus is on the spatial dimension of companies' innovation activities and on local and regional innovation environments.

Christine Liefooghe, Maître de conférences, UFR de géographie et aménagement, Université de Lille 1, Bd Langevin–Cité scientifique, 59650 Villeneuve d'Ascq. E-mail: christine.liefooghe@univ-lille1.fr. Dr Liefooghe's research foci are on the producer services in the restructuring of industrial regions, and in particular on the distance selling in the Lille metropolitan area.

Christian Longhi, Senior Researcher, CRNS (French National Center of Scientific Research), GREDEG, CNRS and University of Nice Sophia Antipolis; 250, Rue A. Einstein, Sophia Antipolis, 06560 Valbonne, France. E-mail: longhi@idefi.cnrs.fr. Dr. Longhi's research focus is on regional economics, local and sectoral systems of innovation and production, ICT and economics of the internet. The sector of tourism constitutes a field of research particularly developed by Dr Longhi.

Sam Ock Park, Dean and Professor, College of Social Sciences, Seoul National University, Seoul, 151–746, Korea. E-mail: parkso@snu.ac.kr. Professor Park's research focus is regional innovation systems, industrial clusters and regional development. He served as President of Pacific Regional Science Conference Organization and the Pacific Editor of *Papers in Regional Science.*

Sang-Chul Park, Professor of E-Business and Energy Policy, Graduate School of Knowledge-based Technology and Energy, Korea Polytechnic University, 2121 Jeongwang-Dong, Siheung-City, Kyonggi-Do, 429-793, Korea. Email: scpark@kpu.ac.kr or scpark86@hotmail.com. Professor Park's research focus is regional development strategy based on innovative clusters. His research interests also include regional policies for building science parks and technology parks, the analysis of technology innovation, and systemic innovation within regions.

Joanne Roberts, Lecturer in International Business, Durham Business School, University of Durham, Mill Hill Lane, Durham DH1 3LB, United Kingdom. E-mail: joanne.roberts@durham.ac.uk. Dr Roberts's research interests include the internationalization of knowledge intensive services and new information and communication technologies and knowledge transfer.

Mercedes Rodríguez, Assistant Professor, Department of Applied Economics, Campus Universitario de Cartuja s/n, E-18071, Granada (Spain). E-mail: m_rodrig@ugr.es. Assistant Professor Mercedes Rodriguez's research focus is innovation in services. More concretely, she studies innovation in knowledge-intensive services (KIS).

Grete Rusten, Senior Researcher, Institute in Economics and Business Administration (SNF) Breiviksveien 40, 5045 Bergen, Norway. E-mail: Grete.Rusten@snf.no. Dr Rusten's research focus is FDI-strategies and regional effects, outsourcing and business services and SMEs. An ongoing project with Bryson at University of Birmingham is about the production and consumption of design services. She is an Honorary Research Fellow at School of Geography, Earth and Environmental Sciences, University of Birmingham and a Council Member of European Research Network on Services and Space (RESER).

Christian Schulz, Professor in Geography and Spatial Planning, University of Luxemburg, 162a Avenue de la Faiencerie, L-1511 Luxemburg, E-mail: christian.schulz@uni.lu. Professor Schulz's research focus is on service industries as well as on environmental issues in economic geography.

Patrik Ström is a post-doc research fellow at the Swedish Collegium for Advanced Study in the Social Sciences (SCASSS) in Uppsala, Sweden and the Department of Human and Economic Geography, School of Business, Economics and Law, Göteborg University in Sweden. Dr Ström can be contacted at the Department of Human and Economic Geography, School of Business, Economics and Law, Göteborg University, P.O Box 630, S – 405 30 Göteborg, Sweden. E-mail: patrik.strom@geography.gu.se. His research focuses on the development and internationalization of the Japanese service industry.

Peter Wood, Professor, Department of Geography, University College London, Gower Street, London WC1E 6BT, United Kingdom. E-mail: p.wood@geog.ucl.ac.uk. Professor Wood's long-term research interests are in economic restructuring and regional development, but his focus in recent years has been on the service sector, especially the growth of producer and business services.

Preface

Several words are used constantly to characterize the contemporary economy, including 'global,' 'service,' and 'knowledge-based.' These adjectives appear everywhere (book titles, journalistic coverage, and even in popular epithets), seldom acknowledging that these characteristics are related. Service activities based on the collection, manipulation, and dissemination of information underpin the operation of global production and consumption networks. Information and communication technologies, in turn, drive some of these reconfigured relationships among producers, suppliers, and consumers – relationships which are usually mediated by knowledge-based services. This book makes some of those relationships explicit.

Services are everywhere in our economy, whether as a set of very large and diverse sectors or as the basis for all economic activity. This is visible in the core activities of some of the world's largest companies, the occupational dimensions of employment growth, and in the sources of innovation and efficiency in global procurement, control, and consumer relationships. Yet what is visible is not necessarily *clear*. What makes an activity 'knowledge based'? What are the implications of services' roles in wresting value-added from knowledge? How do service firms manage the combination of international reach and locally-based knowledge? What are the sources and implications of local variations in the prevalence, structure, and relationships of knowledge-based services?

These are just a few of the questions explored during an international conference held at the University of Birmingham in August 2004. Titled 'Service Worlds: Employment, Organizations, Technologies,' it was the annual residential conference of the International Geographical Union's Commission on the Dynamics of Economic Spaces. Peter Daniels and Michael Taylor served as the organizers and conveners. Thirty-five papers were presented and discussed, with contributors and case examples from East and Southeast Asia, Europe, and North America.

In the succeeding months, we communicated with the presenters, helping them craft a set of chapters that speak to the questions above. Having worked closely with these authors and their contributions for nearly two years, we have grown even more aware of their interrelationships and the practical and theoretical utility of knowledge services in contemporary economies. We are indebted to all the contributors for patiently awaiting our initial decisions about which papers to include in this volume and their subsequent prompt responses to our requests for changes to their manuscripts or the inclusion of additional material. The prompt publication of volumes of this nature relies heavily on the cooperation of the contributors; we gratefully acknowledge their generous support for this project.

James W. Harrington, Jr   Peter W. Daniels
Seattle, WA, US   Birmingham, UK

List of Abbreviations

ABI	Annual Business Inquiry (UK)
ASEAN	Association of Southeast Asian Nations
B2B	business-to-business
B2C	business-to-consumer
BERD	business enterprise research and development
DMO	destination management organization
EC	electronic commerce
EU	European Union
FDI	foreign direct investment
GATS	General Agreement on Trade in Services
GATT	General Agreement on Tariffs and Trade
GDP	Gross Domestic Product
GDS	global distribution system
GSE	Greater South East (England)
IB	international business
ICT	information and communications technologies
IT	information technology
JETRO	Japanese External Trade Organization
KIBS	knowledge intensive business services
KIS	knowledge intensive services
LLS	local labor system
MCS	management consultancy services
MNC	multinational corporation
MSA	metropolitan statistical area
OECD	Organization for Economic Cooperation and Development
OLI	ownership, location, and internalization
PSE	post-secondary education
R&D	research and development
SME	small- or medium-sized enterprise
TMA	Turin metropolitan area
TPS	Turin Production System
UK	United Kingdom
US	United States

Chapter 1

International and Regional Dynamics of Knowledge-Based Services

James W. Harrington and Peter W. Daniels

This volume presents research addressing the organization, international relationships, and regional economic impacts of a broad set of service-providing activities. These activities include the widely studied 'knowledge-intensive business services,' but go beyond these to recognize knowledge creation and dissemination in the post-secondary education sector and the knowledge about markets and value chains that underlies the burgeoning mail-order and e-tourism sectors. While the research proceeds from varied theoretical perspectives, the chapters overall spring from three major premises, explained in turn in the following pages.

1. The acquisition and organization of knowledge (information combined with context to yield useful guidance) are central to the operation and marketing of many service-providing activities.
2. These requirements motivate the organizations' structure, relationships to other organizations, location of operations, and entry into new markets.
3. Because the knowledge requirements vary by service sector and the opportunities for organizations' structure and relationships vary by local context, sector- and location-specific studies are necessary to explore the nature of these contingent relationships.

Information, Knowledge, and Competences: Importance

We read constantly of the 21st century's 'knowledge-based economy: a new historical era where the economy is more strongly and more directly rooted in the production, distribution and use of knowledge than ever before' (Foray and Lundvall 1996, 12). However, economic activity remains quite material, with a major stake in the flow of energy, goods and physical investment. What are the sources of organizations' economic success and sustainability? The record profits made by major oil companies and the economic power of huge companies such as Wal-Mart remind us that the ability to exploit a scarce resource and the ability to exert market power on suppliers remain important. However, the management of knowledge about markets, suppliers, finance and innovation underlies these successful companies, and successful not-for-profit activities (Porter 1990). To quote Freeman (1982, 4) 'the investment process is as much one of the production and distribution of knowledge as the production and use of capital goods, which merely embody the advance of science and technology.'

While Sayer and Walker (1992) shun labels that imply major shifts in the nature of capitalism, they imply quite clearly that they see the proliferation of specialist business service firms as a way of commodifying expertise and making it necessary for successful competition. Therefore the capitalists controlling firms engaged in certain activities are compelled to pay knowledge-service entities a profit above their own cost of producing and disseminating expertise. In their view, the rise of the service economy is just a manifestation of an ever-increasing division of labor, subject to the same logic as the rest of capitalism. The increased commodification of knowledge has led to a spate of corporate entities to help protect and broker 'intellectual property.'

Information, conceptualized and contextualized to serve as useful knowledge, forms the major input, the major product, and a primary strategic asset for many service providers – across private, public, and not-for-profit sectors. Pure knowledge, of course, is of limited utility. The organizational literature focuses on usable knowledge, or competences. 'By "competence,"' Foss (1996, 1) 'understands a typically idiosyncratic knowledge capital that allows its holder to perform activities … in certain ways, and typically do this more efficiently than others.' Foss notes explicitly how much this owes to the strategic approach to studying firms and competition, originating from the Harvard Business School and Michael Porter's emphasis on competitive strategy (see Prahalad and Hamel 1990). More simply, Belussi and Gottardi (2000) distinguish (raw) information, (codified and tacit) knowledge, and (organizational) competence, which is the ability to deploy knowledge to improve production, marketing and competitiveness.

Loasby (1996, 41) outlines two dimensions along which organizations must assess and manage their competences: 'the degree of specificity' of application and 'the degree of control' – how well and how exclusively does the firm control the competence. Note that owning or internalizing a competence is a capital investment that may not bring a return if the competence is so specialized that it's never or seldom needed, or if mere 'ownership' is insufficient for its use (it may need complements; it may need greater expertise in other parts of the organization; it may be embodied in individuals in whom the firm has invested, but who then leave the firm). On the other hand, not owning a competence elicits the costs of identifying, paying, and monitoring an external source of the competence – and even then, internal complements may be required.

Claiming competence is of special importance to service firms' marketing. Clients are forthcoming because the highly differentiated nature of knowledge and expertise relevant to organizations' operations, innovation and competition leads some firms to consult outside expertise. Bryson (1997) suggests two key reasons: to substitute for in-house capability, thereby 'only employing experts when they are required' (100); and to augment in-house expertise, including to gain expertise/insight that may have been used by competitors. 'Business service firms, by operating as innovation transfer agents, may be responsible for the dynamic nature of organizational structures and operational procedures' (101).

Knowledge Acquisition and the Organization of Spaces

Knowledge acquisition and organization motivate the processes through which these providers gain their inputs, interact with other organizations, hire and train their workers, and service their markets. However, inter-firm or intra-firm divisions of knowledge are even more difficult to manage and coordinate than divisions of labor. 'Indeed, the finer the division of labor and of knowledge, the greater the expertise that results, but also the greater is likely to be' the difficulties of coordination, because individuals' or firms' deeper expertise often implies ignorance over a broader area outside their expertise (Loasby 1996, 41).

Innovativeness is a motivation for knowledge-based firms' procurement, hiring, location, production, and marketing. Oerlemans and Meeus (2005) asked managers of 365 Netherlands firms (in manufacturing and service sectors) about the nature of 'external contributions to the innovation process' (98). They specified 12 possibilities, and performed factor analysis to come up with four factors:

F1: 'intermediaries': chambers of commerce; trade organizations
F2: 'educational institutes': general universities; technical universities; colleges; vocational institutes
F3: 'business agents': important buyers; important suppliers; competitors
F4: 'innovation advisors': 'national centre for applied research'; 'innovation centres'; 'consultants.'

They concluded that firms' internal characteristics *and* external linkages are key determinants of innovativeness, and it is insufficient to assume that external linkages for innovation will occur because of regional characteristics. However location and co-location are powerful influences on the intra-organizational and external knowledge of which firms can avail themselves.

From a geographic perspective, these processes influence the paths that knowledge-based service providers take as they attempt to gain footholds in international markets, and the ways in which services providers operate under varied local circumstances. Geographers suggest that repeated interaction (which is made more likely through proximity – but is distinct from proximity) reduces the costs of identifying and monitoring external sources of competence, thereby increasing the feasibility of a fine division of knowledge.

However, for continuous innovation, widespread links may be more important than localized links to information and knowledge. Cooke et al. (2005) surveyed 455 SMEs (of 3600 to whom the survey was mailed), grouped into five sectors, across 36 administrative regions of the UK, across all 12 UK Standard Regions. 'Innovative firms tend to make greater use of collaboration and information

Table 1.1 Information requirements and locational motivations of information-based services

Type of service	Information/knowledge requirements	Geographic reference points	Locational motivation
Back offices	routine information	no need for proximity to clients	cost driven
Interpersonal services	structured information exchange	market based	driven by access to clients and to skilled workers
Highly specialized services	create or apply knowledge in novel situations	skill and interaction based	driven by access to skills and creativity base

exchange, be involved in higher trust relationships, and make greater use of non-local networks ... Overall, our pattern of results suggests that 'firm' effects are more significant than 'regional' effects per se; certain types of firms (e.g., innovators or knowledge businesses) tend to make greater use of particular forms of social capital and, typically, there are greater numbers of these firms in the more favoured regions' (1074).

'Services' are a notoriously heterogeneous set of activities. To increase its coherence, this volume focuses on knowledge-based services. However, not all of these rely on the creating and marketing of totally new knowledge. Table 1.1 augments Illeris's (1994) levels of knowledge intensity among information-intensive services and the implications for location of their activities.

With regard to highly specialized services, Simmie (2005, 798) emphasizes the importance of (a) information exchange and diffusion by way of international service firms, and (b) the concentration of these firms' major offices in the largest cities in wealthy countries, motivated by the location of both key clients and key sources of highly trained personnel. 'Because such workers tend to "stick" to their regional labor markets, this raises the propensity for innovative activities to concentrate in the same region throughout all phases of their life cycles (Audretsch and Feldman 1996).' However knowledge-based and highly specialized services appear in every type of region, in nearly every country. The following chapters portray this variety of locations, locational change and economic impact.

Purposes and Organization of this Volume

This book is motivated by a need to move beyond the study of 'knowledge-intensive business services' to a broader set of organizations that provide services based on knowledge. As noted above, it has three major premises:

1. Knowledge acquisition and organization are central to the operation and marketing of many service-providing organizations.
2. These requirements motivate the organizations' structure, relationships to other organizations, location of operations, and entry into new markets.
3. Because the nature of knowledge requirements vary by service sector and the opportunities for organizations' structure and relationships vary by local context, sector- and location-specific studies are necessary to explore the nature of these contingent relationships.

This book offers theoretical and empirical insights into these processes and paths. Each author ultimately emphasizes the special attributes of knowledge acquisition and diffusion within and across organizations, and the consequent roles that these structurally important firms and institutions play in regional economic development. Conceptually, the authors rely on and add to industrial organization, spatial interaction, transactions costs and regional economic structure/structural change.

Part 1

The remainder of the volume comprises 13 chapters in 3 sections. The chapters in Part 1: 'Conceptualising Knowledge-Based Services,' describe what these activities have in common, how they obtain and develop knowledge, and what roles they play in broader systems of innovation.

In 'Service Worlds and the Dynamics of Economic Spaces,' Sam Ock Park (Seoul National University) considers spatial interaction in the context of advanced services, and the consequent dynamics of economic spaces. Transferability, knowledge-genesis, networks, collaboration, and hierarchy of control are suggested as key determinants of spatial interaction among knowledge-based activities. Intensified spatial divisions of labor, clustering of advanced services, internet-based services, globalized networks of services, and virtual innovation clusters are identified as the major influences on the dynamics of economic spaces in service worlds. In-depth surveys regarding electronic commerce by Samsung Electronics and innovation networks in the Sunchang region support the importance of these interaction determinants. The dynamic integration of these determinants of interaction in the knowledge-based information society has resulted in the dynamic and controversial economic spaces in core and peripheral areas such as clusters and global networks; intensified spatial division of labor and virtual innovation clusters. Park concludes by drawing policy implications for developing areas or peripheral regions.

In 'Knowledge Intensive Services and R&D Diffusion: An Input-Output Approach,' José Camacho and Mercedes Rodríguez (University of Granada) start from the premise that service activities are dramatically increasing their participation in the production processes and, as a result, they account for the bulk of the value added in most industries. Within this context, Camacho and Rodriguez calculate the embodied R&D flows that are generated in the supplying of services in six European countries: Denmark, Finland, France, Germany, Spain and the United Kingdom. The results obtained demonstrate the strategic role that services, and in particular knowledge-based services, play in innovation diffusion.

In 'Innovation and Technological Change in Tourism: A Global-Local Nexus,' Christian Longhi (CNRS and University of Nice Sophia Antipolis) focuses on the ways in which technological and organizational innovations have dramatically altered the organization of and markets for tourism. The tourism sector has not only been able to absorb the knowledge bases and technological changes of related provider sectors, but has worked as an engine of change with strong capabilities for innovation. E-tourism encompasses a huge share of e-commerce, and several related technologies have been developed by the main actors in the tourism sector. This chapter's study of tourism services underlines the interdependence among business models, organizational forms, and technologies. Internet-based and -enabled technologies have shifted the locus of power within supply chains, led to new corporate forms and combinations, and have affected the ability of localities to create and market tourist experiences. Europe is a center of tourist activity and of the changes that have developed. By establishing a framework for the study of structural change in tourism, Longhi deciphers the changes in Europe to date and makes some forecasts for the future.

Part 2

The chapters in Part 2 explore processes and implications of the internationalization of services. How do firms meet the exigencies of developing and customizing knowledge relevant to varied national contexts? Several of these chapters echo the arguments above, that organizational and inter-organizational networks can be more important than localization, for purposes of sharing core knowledge.

In 'Spatial Divisions of Expertise and Transnational "Service" Firms,' John Bryson (University of Birmingham) and Grete Rusten (Institute for Research in Economics and Business Administration, Bergen) use the literature on global cities to suggest that the most valuable and 'world-class' expertise is to be found in the global cities; these are the command and control centers of global capitalism. Yet this literature does not address to any great extent the ways in which transnational service companies combine expertise developed and maintained in a variety of different geographical contexts, many of which lie beyond the global cities. This chapter explores the organizational spaces of transnational professional service firms, drawing empirical cases from the aerospace and management consultancy sectors. These firms are developing and exploiting a complicated functional and spatial division of labor/expertise in which local expertise is increasingly mobilized globally. Such mobilization is developed by creating virtual and actual spaces of social interaction that are intended to ensure that transnational service firms utilize the value of their knowledge base. The approach developed in this paper combines perspectives developed in the literatures on global cities and transnational corporations by exploring the spatial divisions of expertise that exist within global service suppliers. Global cities may appear to be the major locations for the provision of knowledge intensive services, but, in many instances, clients and experts meet and work together outside these cities. The point is that there are other spaces of interaction (physical, social virtual, organizational) that lie beyond the global cities and that increasingly understanding the operation of the global knowledge community implies that geographers focus on addressing the organization of production and the ways in which this interacts with a variety of different places.

In 'Internationalization of Management Consultancy Services: Conceptual Issues Concerning the Cross-Border Delivery of Knowledge Intensive Services,' Joanne Roberts (University of Durham) explores international management consultancy services in order to identify conceptual issues concerning the cross-border delivery of knowledge-intensive services. A definition of management consultancy services and an account of their growth and development is followed by a discussion of the internationalization of UK management consultancy services, and a review of the cross-border transfer of knowledge. The role of management consultancies in the cross-border transfer of knowledge is then considered, to construct a framework that illustrates the various mechanisms through which knowledge transfer occurs in the cross-border delivery of management consultancy services. The chapter concludes with an agenda for further research.

A plethora of late twentieth century studies focused on the role of services in economic development. Case studies reported that internationalization was driven by the desire to deliver bespoke products to clients through international offices so that quality control, reputation and trust would remain internalized within the firm. In 'The Internationalization of Europe's Contemporary Transnational Executive Search Industry,' Jonathan V. Beaverstock and Sarah J.E. Hall (Loughborough University) and James R. Faulconbridge (Lancaster University) study elite executive search and selection to investigate the contemporary globalization of a transnational service industry whose success is built upon the firm's knowledge-rich workforce, trust relations with clients, reputation and flexible modes of strategic operations. Headhunting emerged as a global industry from the US and Europe as a consequence of the global spread of FDI by the key headhunting firms who were striving to become TNCs, coming to age during the 1980s 'boom' in cities like Chicago, New York, London, Paris and Brussels. Like other producer services, headhunting has become a highly concentrated industry. After the US, Europe is the most significant market for the activities and locational dynamics of headhunters in the world, and therefore, it is Europe that is the object of the study. This chapter unpacks the strategic importance of headhunting as a key European industry for the recruitment of executive labor, and sheds theoretical light on the complex processes which drive globalization and deepening of transnational service firms in a mature market. Drawing upon extensive data on headhunting firms in Europe, we trace the longitudinal nature of their growth and changing locational dynamics. Finally, the authors comment on the challenges facing headhunting firms in Europe as they restructure operations, for example from wholly owned models to members of strategic alliances, to penetrate new markets and remain competitive in a very concentrated sector of the world economy.

Among OECD countries, Japan's economy has a relatively small share of the GDP and employment that is generated by the tertiary sector. Additionally, Japan runs large trade deficits in service trade and has a very low export ratio to total service production. In the international service industry few Japanese multinationals are more competitive than most of the dominating Western firms. Why does it seem that Japan lags behind in terms of service industry development and internationalization? Have the specific organizational drivers within the Japanese business environment influenced the international competitiveness of Japanese service firms? In 'Internationalization of Japanese Professional Business Service Firms: Dynamics of Competitiveness through Urban Localization in Southeast Asia,' Patrik Ström (Göteborg University) pursues these questions through a study of Japanese professional business service firms active in Japan and Singapore. The facts that Asia over the last three decades has seen tremendous economic growth and that Japan has played an important part in this regional development should create good business opportunities for Japanese business service firms. The relative geographical closeness to many of the rapidly growing economies in Asia along with the large inflow of Japanese foreign direct investment could act as a ground for a better position of Japanese firms than the highly competitive business service markets in North America and Europe. By using a case-study method with in-depth interviews in ten firms, Strom explains why and how Japanese firms have chosen to locate in Singapore in relation to their natural Asian presence through the Japanese head office. The results reveal the importance of Japanese related business and how the complex ownership structures within the *keiretsu* groups affect the regional competitiveness. The results can further be used in comparative research on Western market leading firms, which are active competitors even in the domestic Japanese market.

Part 3

Insofar as useable knowledge about production, organization, markets, and finance is the basis for commercial, non-commercial, and governmental success, services that create and provide knowledge are key to the fortunes of nations and regions. 'Business service companies are increasingly driving innovations in the wider production system. They provide expertise that represents the accumulation of knowledge from working for a variety of different companies and in different countries' (Bryson 1997: 108). Part 3, 'Knowledge-Based Services and Regional Development' studies these aspects of knowledge-based services, emphasizing the variation in regions and service sectors.

The transformation of regional economies from manufacturing towards services relies on varied structural mechanisms that act on specific sectors. In addition, these development mechanisms interact with the different institutional, social, and cultural histories, giving origin to distinct models of post-industrial capitalism. In 'Towards Post-Industrial Transition and Services Society? Evidence from Turin,' Paolo Giaccaria and Vincenzo Demetrio (University of Turin) undertake an analysis of structural and institutional change in the Turin region. They suggest that its transition towards a service economy presents some similarities with the neo-industrial model as defined in the literature, but with some peculiar differences. Survey analysis suggests that there is a strong commitment of service firms towards establishing cooperative networks within the region. This is particularly evident in the relationships of service related firms to innovation and research facilities within the metropolitan area.

Post-secondary education (PSE) has been increasingly emphasized in studies of and policies for economic growth and development. In 'Post-Secondary Education: Education, Training and Technology Services,' Hyungjoo Kim (Korea Science and Technology Policy Institute and Sungshin Women's University) and James Harrington (University of Washington) report on a study of IT-producing firms' use of and interaction with PSE institutions. How distinct are the types, degree, and geography of services provided by different types of PSE institutions? Does geographic distance inhibit different types of interaction differently? Do firms' type and degree of interaction vary by size, sector, or region? While most research on relationships between higher education and industry focus on research-oriented, doctoral-degree-granting universities, this research includes baccalaureate colleges, community colleges, and technical colleges. The authors selected three MSAs based on characteristics of IT-producing industries and universities/colleges located in each area: Seattle, WA; Portland, OR; and Champaign-Urbana, IL. Eighty IT-producing establishments responded to the on-line survey questionnaire between January and May 2004. Overall, the research findings support the oft-cited assertions that universities support economic development, and that this support is fairly localized. The findings suggest that lower- and non-degree-granting institutions (community, junior, community, and four-year colleges) play large roles, as do the doctoral and research universities more often studied in the academic economic development literature.

In 'Danish Regional Growth Strategy in Marginal Areas: Regional Partnership and Initiative,' Sang-Chul Park (Okayama University) explores how the Danish government has established policies to strengthen its knowledge-based economy and to carry out knowledge-based regional growth strategies. Knowledge-based economic growth affects the nature of regional economic growth and the interregional dynamics of growth, which are key policy concerns of the government. A unique strategy of developing niche technologies, and a growth strategy for marginal areas, have contributed to positive growth rates. However Denmark faces limitations in its technology industries, in part because its primarily small firms have been reluctant to carry out R&D activities.

In 'Urban Revival and Knowledge-Intensive Services: The Case of the English "Core Cities,"' Peter Wood (University College London) examines the representation of KIBS in these 'core' English cities and their future prospects as nodes in the networked, knowledge-based economy. The growth of tradable, expert knowledge intensive business services (KIBS) is a largely urban phenomenon, favoring larger cities and their regions. This has important implications for urban economic development, not just because of the growth and quality of KIBS employment itself, but also its influence on the success of other sectors. The regional development of KIBS reflects demands from other service, manufacturing, and trading functions, MNCs and other large firms, growth-orientated SMEs, regionally distinctive consumer and cultural sectors, and public agencies. The level and range of KIBS supply depends on a regional 'critical mass' of such specialist demands, as well as regional institutional and regulatory support. UK cities outside London face much greater challenges from the KIBS dominance of the capital than do comparable cities elsewhere in Western Europe. They thus need especially effective strategies to promote the coherent development of tradable knowledge-based assets, extending across both the commercial and public sectors.

In 'Innovation Activities of KIBS Companies and Spatial Proximity: Some Empirical Findings from Finnish New Media and Software Companies,' Jari Kolehmainen (University of Tampere) notes the many roles that knowledge-intensive business services (KIBS) play in the current economy. Innovation studies have suggested that the innovation activities of service companies differ from traditional industrial innovation. In addition, studies highlight the importance of companies' external networks within innovation activities and the role of continuous learning and incremental innovations. On the basis of this new 'recursive' innovation model, many authors suggest that an appropriate local operational environment can have a positive effect on companies' possibilities to innovate. To bring some clarity to this very complex and intriguing issue, Kolehmainen introduces the concept of local innovation environment. He argues that a local innovation environment consists of many interrelated elements ranging from the institutional setting to the behavior of individuals. It is also argued that the local innovation environment can be divided into three levels: the structural and institutional level; the level of organizational relationships; and the level of individuals. These ideas were empirically examined from the perspectives of 105 Finnish new media and software companies. The empirical results parallel those of previous studies.

Christian Schulz (University of Cologne), H. Peter Dörrenbächer (University of the Saarland), and Christine Liefooghe (Universite des Sciences et Technologies de Lille) collaborated to write 'Far Away, So Close? Regional Clustering of Mail Order Firms and Related Business Services in the Lille Metropolitan Area.' After the decline of traditional industries such as textile, iron and steel as well as coal mining, the regional economy of Nord-Pas-de-Calais/France has undergone a profound restructuring process in which advanced business services play a crucial role. One of the most remarkable evolutions is the formation of a highly specialized service cluster due to the presence of France's major mail order houses, either succeeding or being founded by former textile companies. A large variety of knowledge intensive business services (e.g. logistics firms, PR consultants, designers and product photographers) constitute a localized network and apparently represent an indispensable environment for mail order firms. This chapter approaches this topic in two ways. First, a theoretical framework based on evolutionary approaches is discussed in order to conceptualize the emergence of these particular activities in an old industrial region. Second, results of an exploratory study are presented in order to document characteristic path dependencies and their impacts on the services sector in Nord-Pas-de-Calais. In addition, current tendencies and future prospects are discussed – not least with regard to the growing importance of internet retailing and e-commerce.

Even within a narrowly defined service sector, organizations take on specific markets and roles. Despite the variety of organizations, sectors, and regions, these chapters make clear the importance of inter-organizational linkages within and among regions, for obtaining and marketing useful knowledge.

References

Audretsch, D.B. (1998), 'Agglomeration and the Location of Innovative Activity', *Oxford Review of Economic Policy* 14: 2, 18–29.

Audretsch, D.B. and Feldman, M.A. (1996), 'Spillovers and the Geography of Innovation and Production', *American Economic Review* 86, 630–640.

Belussi, F. and Gottardi, G. (eds) (2000), *Evolutionary Patterns of Local Industrial Systems: Towards a Cognitive Approach to the Industrial District* (Aldershot, England: Ashgate).

Belussi, F. and Gottardi, G. (2000), 'Models of Localised Technological Change', in Fiorenza Belussi and Giorgio Gottardi (eds).

Bryson, J.R. (1997), 'Business Service Firms, Service Space and the Management of Change', *Entrepreneurship and Regional Development* 9, 93–111.

Cooke, P., Clifton, N., and Oleaga, M. (2005), 'Social Capital, Firm Embeddedness and Regional Development', *Regional Studies* 39: 8, 1065–1077.

Foray, D. and Lundvall, B.Å. (1996), 'The Knowledge-Based Economy: From the Economics of Knowledge to the Learning Economy', *Employment and Growth in the Knowledge-Based Economy* (Paris: OECD).

Foss, N.J. and Knudsen, C. (eds) (1996), *Towards a Competence Theory of the Firm* (London: Routledge).

Freeman, C. (1982), *The Economics of Industrial Innovation*, sec. ed. (London: Pinter).

Illeris, S. (1994), 'Proximity Between Service Providers and Service Users', *Tijdschrift voor Economische en Sociale Geografie* 85, 294–302.

Loasby, B. (1996), 'The Organization of Industry', in Nicolai J. Foss and Christian Knudsen (eds).

Oerlemans, L.A.G. and Meeus, M.T.H. (2005), 'Do Organizational and Spatial Proximity Impact on Firm Performance?' *Regional Studies* 39: 1, 89–104.

Porter, M. (1990), *The Competitive Advantage of Nations* (New York: Free Press).

Prahalad, C.K. and Hamel, G. (1990), 'The Core Competence of the Corporation,' *Harvard Business Review* 68: 3, 79–91.

Sayer, A. and Walker, R.A. (1992), *The New Social Economy: Reworking the Division of Labor* (Cambridge, Mass.: Blackwell).

Simmie, J. (2005), 'Innovation and Space: A Critical Review of the Literature', *Regional Studies* 39: 6, 789–804.

PART 1
Conceptualizing
Knowledge-Based Services

Chapter 2

Service Worlds and the Dynamics of Economic Spaces

Sam Ock Park

Information and communication technology (ICT) has transformed production systems and the role of space and time in the pattern of interaction in the space economy. The economic consequences of the revolution in computational and communications capabilities and the myriad of applications pervade every aspect of our lives and business activities (Beyers 2002). The phrases 'digital economy' and 'new economy' refer to the transformation of economies by the application and development of ICT. The new economy is being shaped not only by the development and diffusion of computer hardware and software, but also by much cheaper and rapidly increasing electronic connectivity in the global society (ESA 2000). As part of the new economy, the service world is evolving through the development of new forms of production and consumption mediated by new technology, institutions, laws, and new division of labor (Bryson et al. 2004). Services now permeate every part of contemporary economy and society, affecting the dynamics of economic spaces. The service-mediated economy inextricably combines information, services, movement, and goods.

Advances in ICT and globalization have prompted significant changes in production and consumption. Firms have introduced new forms of management, organization and work in response to the forces of competition. Symptomatic of these changes is the transformation of employment away from manufacturing to services. In advanced economies, the vast majority (often more than three quarters) of all jobs involve services of one form or another. Furthermore, new jobs are much more concentrated in services in the economically developed countries (Bryson et al. 2004). This shift does not necessarily imply that all economies are moving away from manufacturing into services, but there is an ongoing 'blurring' of the long established distinction between the two. This on-going transformation has stimulated new support functions that feed into the production process (manufacturing and services) as well as increasingly driving it. This is an important adjustment to the operation of the capitalist system. Meanwhile, the less tangible aspects of the production process are performing an ever more important role in the design, production and sales of goods and services. A new economic geography that incorporates production and consumption is developing in which service industries are amongst the key players. They are also important in promoting local economic growth in different national settings, with substantial and different impacts in developed and developing countries.

Interactions among the producers of information, knowledge, services, goods, manpower, and strategies have become more diversified in terms of intensity, scope, and spatial scales. The influence of space and place on these interactions differs from our traditional understanding of spatial interaction. This chapter reviews the notions of interaction in the advanced services and conceptualizes the dynamics of economic spaces in a knowledge-based economy. The chapter ends with examples of changing spatial interaction and organizational dynamics in Korea.

Notions of Interaction in the Service Worlds

Interaction in knowledge-based economies differs from our notions of commodity-based interaction. Traditional notions of spatial interactions focused on flows of commodities and people. Ullman's (1957) triad of transferability, complementarity, and intervening opportunity is considered the classic basis for spatial interaction. Transferability is related to the cost of movement and is the basis for classic location theory; complementarity reflects the spatial inequalities of supply and demand of commodities; and intervening opportunity considers spatial competition and substitutability among destinations. Wheeler and Mitchelson (1989) examined three hypotheses regarding to information flows among metropolitan areas in the United States: information genesis, hierarchy of control, and spatial interdependences. Based on the information flows among 48 large metropolitan areas in the United States using Federal Express Corporation data, they supported the three hypotheses. That is, 'supply considerations, rather than demand, are fundamental in information genesis; flows are strongly asymmetrical, reflecting a marked hierarchy of control; and distance plays a minor role in the spatial configuration of flows, especially at the highest level of the metropolitan hierarchy' (Wheeler and Mitchelson 1989, 523). The findings suggest that knowledge-intensive economic activities respond to different bases for spatial interaction and integration than do commodities.

Ullman's triad is still applicable to ordinary commodities. However, the spatial range of knowledge-intensive services has significantly increased in the Internet era. The growth of many producer services is closely related to the externalization of the services and increasing complexity of the division of labor. Outsourcing is now common among large service firms. Because of the increasing externalization trend of the services, knowledge-intensive services have different bases of spatial interaction. In the service economy, knowledge and knowledge-based services are interwoven at all levels of production processes. As seen in Figure 2.1, the production process, from production to consumption, 'involves the articulation of various forms of tacit and explicit knowledge, raw material, land, building and people' (Bryson et al. 2004, 52). In the production processes of pre-manufacturing, manufacturing, distribution and consumption, various services including knowledge-intensive producer services have critical roles in adding value and competitiveness. Tacit and

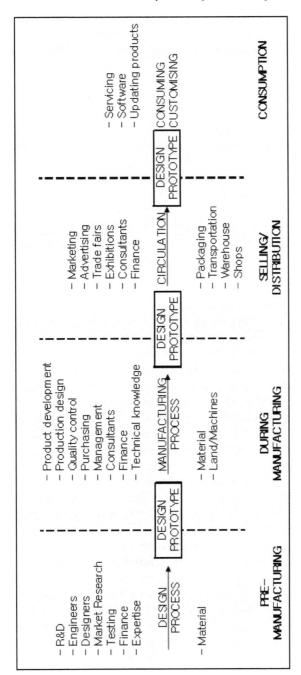

Figure 2.1   Service functions in the production process
Source: Revised from Bryson et al. 2004, 52.

codified knowledge are interwoven, converged, and created during these processes. Four notions of interaction of the advanced services can be identified in the Internet era. They are transferability, knowledge-genesis for service products, network and collaboration, and hierarchy of control.

First, transferability is still important in the service economy. However, the spatial range of advanced services is extended because the impact of distance has decreased significantly for both tangible goods and intangible services. A considerable portion of advanced services can be supplied electronically, representing an overwhelming extension of spatial range. Continuous increases in inter-regional flows of services in the major cities in the Asia Pacific reveal the development of the Asia Pacific service industries within the wider context of trends in global service economy (Daniels 2005). Technological innovation has enabled a number of Asia-Pacific countries to 'leap-frog' development in services. However, ICT can have not only dispersion possibilities but also a centralizing effect in the supply of services, since transferability is important for some activities, but not significant for other activities. In addition, space itself is not smooth, because of the digital divide among countries, regions, and social groups (Leinbach and Brunn 2001; Zook 2002).

Second, knowledge genesis for service products is critical for inter-regional and international interaction of advanced services in the knowledge-based economy. Knowledge creation and innovation for services have become more important for regional competitiveness and development in the knowledge-based information society (Clark and Lloyd 2000; Glaeser et al. 2001). The clustering of talent, innovation and advanced services is closely related to the process of knowledge creation and transfer. Talent or human capital is at the core of generating new knowledge and innovation. Only a few cities or regions in a nation take a role as centers of creativity, knowledge, and innovation for services. Evolution of global financial centers and world cities in the Internet era support the uneven distribution of knowledge genesis in the global space economy. Large metropolitan areas function as command and control points in generating greater level of information than they receive (Wheeler and Mitchelson 1989). Accordingly, it is natural that fundamental asymmetry in the flow of knowledge and talent among cities and regions. Due to the unevenness of the knowledge genesis over space, regional disparity is persistent even in the Internet era.

Third, networks and collaboration are important mechanisms for the generation of new knowledge and innovation for economic activities. Formal and informal meetings have been regarded as important process for generation of new knowledge and information. Networking and collaboration with customers, suppliers, universities, public institutions, and research centers as well as with different functions within a firm are critical for development of new product services. This is why advanced services are concentrated in a few metropolitan areas. Development of ICT has contributed significantly to overcome the distance decay effect in the interaction of codified knowledge. However, telecommunications are still poor substitutes for face-to-face contacts and transferring tacit knowledge. The development of ICT has made codified knowledge more ubiquitous, but tacit knowledge is still place-base and locally embedded (Park 2003; 2004). Due to the importance of network and collaboration, clustering of advanced services can be continuously progressed unevenly in the global space economy.

Fourth, hierarchy of control in the flow of advance services is important because distance will play an insignificant role in the flow of advanced services. The hierarchy of control is related with the integration of the above three notions: transferability, knowledge genesis, and network and collaboration. The range of advanced services has been overwhelmingly extended in the 21st century due to the liberalization of advanced services and development of ICT in the global economic spaces, representing the increase of transferability. A greater service center is usually the center of knowledge creation with strong network and collaboration. Global financial activities are organized hierarchically with global primary financial centers of New York, London and Tokyo. Advanced financial products and services are produced at a particular point in space even if financial trading is an increasingly ubiquitous virtual activity (Clark 2002). Standardized financial products are more dispersed, while advanced and sophisticated financial services and products can be served only a few leading financial centers. Location of headquarters and regional centers of multinational firms also represents the hierarchy of control in the global space economy. The impact of distance decay has decreased significantly in knowledge-intensive services such as control functions, R&D, design, advertising, business consulting etc. The center of advanced services is organized hierarchically in the order of global, national, and regional centers.

Dynamics of Economic Spaces of Service Worlds

These four notions of interaction interact to influence flows and interaction in the global space economy. Economic spaces evolve dynamically because centripetal and centrifugal forces are coexistent in the process of the restructuring of economy. In this section, four patterns are identified as major dynamic spatial characteristics of service worlds.

Intensified Spatial Division of Labor

It is well known that spatial divisions of labor increased during the industrialization process. There are two basic types of spatial division of labor: sectoral and functional divisions (Massey 1984).

Sectoral divisions of labor develop when regions are specialized in particular industries and related skills. This was a dominant feature of the intense industrialization experienced in Europe and the USA in the 19th century. The development of industrial districts in which particular industries localized within limited areas is a good example of a sectoral spatial division of labor.

Functional divisions of labor occur when firms choose to locate different tasks and occupations within individual industry in different places. As industrial development progressed, different functions within individual industry such as management, marketing, R&D activity, and manufacturing process developed different location factors. The international or interregional product life cycle reflects a dynamic functional division of labor, as standardized production is dispersed to low cost locations (Vernon 1966; Park and Wheeler 1983). In addition, firms may strategically separate well paid managerial or R&D functions from low paid production occupations in order to reduce the bargaining power of low paid production workers as Clark (1981) identified. However, because of the increased need for flexibility in the production process, such spatial divisions of labor can be problematic (Hayter 1997).

In the Internet era, the spatial division of labor has been intensified in the advanced services. There has been a spatial division of labor between R&D activity and production. In addition to this functional division, basic and applied researches tend to concentrate in a core area, while production development tends to disperse to peripheral areas (Park, 1993). Within the R&D activities, more complex and basic research which requires higher quality manpower tends to concentrate in a few major metropolitan areas. This tendency of spatial separation within the same function have intensified the existing functional spatial division of labor, suggesting that regional disparity has not been reduced in the Internet era.

Clustering of Advanced Services and the Internet Industry

Clustering of economic activities is one of the most prominent phenomena in the knowledge-based economy. Clustering of the software industry, venture capital, advanced financial services, producer services, etc. can be clearly identified in the global space economy. Even in the industrializing and newly industrialized countries the clustering of the advanced services are prominent as seen in Bangalore in India, Beijing in China, Seoul in Korea, and the city-states of Hong Kong and Singapore (Park, 2005; Yeh, 2005).

In addition to the advanced services, Internet industries are also clustered in a few places. Despite the Internet's ability to transcend space, dot-com companies clustered in a few major metropolitan areas (Zook, 2005). In Korea, more than three quarters of business-to-consumer (B2C) E-market places, business-to-business (B2B) E-commerce sites, and Internet domains are clustered in the Capital region (Park, 2004). Supplies of skilled labor, the organization of labor, government-support programs, and regional innovation capacity, etc. can be regarded as factors behind the geographic clustering (Zook, 2005).

However, to understand important underlying factors for the clustering, the process of knowledge generation should be noted. Nonaka and Takeuchi (1995) suggested that innovation should be understood as a cycle of interaction between tacit knowledge and codified knowledge. Proximity does not matter in transferring codified knowledge in the Internet era since the codified knowledge can be transferred through Internet globally with low cost. However, the transfer of tacit knowledge can take place at a local level where firms share same values, background and understanding of technical and commercial problems (Maskell and Malmberg, 1999). Geographical and cultural proximity provides the access to local relational networks for the transfer of tacit knowledge (Park, 2003). Accordingly, proximity and institution are also important underlying factors for clustering of knowledge-intensive economic activities.

Globalized Networks of Services

Traditionally most services were supplied locally, at whatever range provided an economic market, as codified in central place theory (Christaller, 1966; Lösch, 1954). As service worlds have become more sectorally and functionally specialized, producer services such as finance, legal, R&D, advertising, technical, and engineering services have increased their international reach. International trade of producer services has significantly increased during the last decade in the Asia-Pacific Cities (Daniels, 2005). Global commodity chains and related services reflect the importance of global networks of services in the knowledge-based economy (Gereffi and Korzeniewicz, 1994).

The global networks of services are more prominent in the B2C electronic commerce such as Amazon.com. In recent years many global firms increased the ratio of global sourcing through B2B electronic commerce (Timmers, 2000; Park, 2004). There are clear hierarchies of urban centers supplying the services. Distribution of Internet firms, domains, and web sites shows concentration of the supply of the Internet related services in major metropolitan areas (Zook, 2005). The distribution of the Internet industry confirms that globalized networks of services intensified in the Internet era on the one hand, with localization of the Internet related service firms on the other hand. Accordingly, the globalization and localization processes are simultaneous components of service worlds.

Virtual Innovation Clusters

In the Internet era, virtual space or cyber space represents another dimension of communication and social space. In the virtual space flows of information and knowledge take place along the networks of Internet. Along the networks, some places take the role of coordinating a smooth interaction in the network to be communication hubs, while other places can be nodes of the networks. Nodes and hubs are hierarchically organized according to their relative importance of the networks (Castells, 2000). Diverse types of innovation clusters are organized in the nodes and hubs, reflecting the spatial division of labor within the value chains of economic activities (Park, 2003).

In the cyberspace cluster of innovation can be organized by networking diverse manpower. In this case hubs and nodes are not necessary condition for the formation of virtual innovation cluster. In the hubs and nodes of real place, actual face-to-face contacts take place easily, accompanying transfer of tacit knowledge and knowledge conversion processes among the high quality manpower such as engineers and information and managerial elites. In the peripheral areas, daily networks of the qualified labors with face-to-face contacts are not easy. However, combining on- and off-line meetings, the cluster of information and knowledge can be possible in a certain place. Research group can be organized through Internet and sets up periodic face-to-face meeting of the advanced researchers to transfer the tacit knowledge in peripheral areas. In this way, even though there is no actual cluster of innovation manpower, cluster of innovation can be formed through combining both on- and off-line networks.

Two Korean Cases

To illustrate and clarify these notions of interaction within the dynamics of service worlds, two contrasting cases have been analyzed based on in-depth interview and field surveys. These cases examine networks through B2B electronic commerce (EC), and virtual innovation networks in a peripheral area. The B2B EC is the format for internet-based transactions between firms. The spatial pattern of Internet industry and B2B EC can illustrate the dynamics of economic space in service worlds. Toward this end, we examine the spatial distribution of .kr domain, firms operating B2B EC, and the spatial networks of Samsung Electronic Company. Secondly, we examine the integration of local culture and resources into innovation networks in Korea's Suchchang county. The two cases will show how the above notions of spatial interaction help explain patterns in the dynamics of economic spaces in the knowledge- and service-based economy.

Spatial Networks of Business-to-Business Electronic Commerce (B2B EC)

Development of B2B EC in Korea. Korea has experienced a rapid growth of the sales of B2B EC thus far in the 21st century. EC sales were 57,558 billion Won in 2000 and represented 4.5 per cent of total sales in Korea. The respective values increased to 177,810 billion Won and 12.2 per cent in 2002, and 314,079 billion Won and 19.3 per cent in 2004. The rapid increase in the sales of EC and their share of total sales seems to be related to the improvement of Internet infrastructure and government policy on the development of EC.

B2B EC shares a dominant portion of the total electronic commerce sales in Korea. In 2001, the percentage of B2B EC sales out of total EC sales was 91.6 per cent. Even though there is a slightly decreasing trend of the share of the B2B EC to total EC sales and an increasing trend of the share of Business-to-Government (B2G) EC, the dominance of the B2B EC sales in the total sales of transaction in Korea has been continued (Table 2.1).

Generally, B2B EC models can be divided into "company-centric models (private B2B marketplaces)" of one-to-many or many-to-one mode and "public B2B marketplaces" of many-to-many mode (Turban et al. 2002). Private B2B marketplaces can be further divided into sell-side (seller-centric) and buy-side (buyer-centric) B2B

Table 2.1 Sales of EC by types of customers

Unit: Billion Won (per cent)

Types	2001	2002	2004
Total	118,976 (100.0)	177,810 (100.0)	314,079 (100.0)
B2B EC[a]	108,941 (91.6)	155,707 (87.6)	279,399 (89.0)
B2G EC[b]	7,037 (5.9)	16,632 (9.4)	27,349 (8.7)
B2C EC[b]	2,580 (2.2)	5,043 (2.8)	6,443 (2.0)
Others	418 (0.3)	427 (0.3)	888 (0.3)

Note: [a] Sales of B2B EC are the sum-up of the sales reported from the B2B EC Firms Survey and the B2B sales surveyed from cyber shopping malls.
[b] Based on Cyber Shopping Mall Survey and B2G EC Survey.
Source: KNSO (Korean National Statistical Office), 2005 (www.nso.go.kr).

Table 2.2 Sales of B2B EC by types of marketplaces

Unit: Billion Won (per cent)

Types	2001	2002	2003	2004
Total	08,941 (100.0)	155,707 (100.0)	206,854 (100.0)	279,399 (100.0)
Buyer[a]	83,167 (76.3)	113,254 (72.7)	150,688 (72.8)	197,212 (70.6)
(Open)[b]		(20.6)	(22.7)	(22.2)
(Closed)[b]		(79.4)	(77.3)	(77.8)
Seller[a]	21,992 (20.2)	36,509 (23.5)	48,766 (23.6)	71,619 (25.6)
(Open)[b]		(12.1)	(12.9)	(11.8)
(Closed)[b]		(87.9)	(87.1)	(88.2)
Public[c]	3,782 (3.5)	5,944 (3.8)	7,400 (3.6)	10,568 (3.8)

Note:
- "Buyer[a]" means buy-side marketplaces (one-to-many); "Seller[a]" means Sell-side marketplaces (many-to-one).
- "(Open)[b]" means B2B EC in which any firm can participate as a buyer or as a seller based on competitiveness;
- "(Closed)[b]" means B2B EC in which participating firms to B2B EC have had a long-term relationships with the operating firms of the B2B EC sites even before the launch of online transactions.
- "Public[c]" means intermediary-oriented e-marketplaces in which many sellers and many buyers can access for transactions.
Source: KNSO (Korean National Statistical Office), 2005 (www.nso.go.kr).

marketplaces. In the buyer-centric type, many suppliers conduct transaction through the EC sites established by a buyer firm, while in the seller-centric type many buyers conduct transaction through the EC sites operated by a single supplier or seller firm for procurement (KNSO, 2005). In private B2B EC, a company that operates the EC site completely controls the participants of EC activities.

Private B2B EC shares more than 95 per cent of the total sales of B2B EC in Korea (Table 2.2). Buyer-centric B2B EC accounts for almost three quarters of the total sales of B2B EC, and seller-centric B2B EC share from 20 per cent to 25 per cent. There is a slight increasing trend in the share of seller-centric type. The public B2B e-marketplaces share is only 3.5 per cent to 3.8 per cent. Considering the dominant share of private B2B EC in Korea, this study focuses on an example of spatial analysis of the company-centric model of B2B EC.

Spatial concentration of the .kr domain and B2B EC in Korea. The internet infrastructure in Korea has become well developed during the last few years. Based on a survey on the number of Internet users and internet behavior in Korea, the number of Internet users was 31.58 million and the Internet usage rate was 70.2 per cent at the end of 2004 (KRNIC 2005). Given that the numbers were 9.43 million and 22.4 per cent, respectively, in October of 1999, the number of Internet users and the usage rate has been rapidly increased during the last five years. In general the younger generation shows a much higher usage rate than the older generations, and the usage rate of males is somewhat higher than that of females. The rate of Internet usage is not significantly different by provincial regions, suggesting that the distribution of Internet users has a similar pattern to that of the population distribution (Park 2004).

However, there is a considerable difference in the usage rate between urban and rural areas. The average usage rates of large metropolitan areas, small and medium cities, and rural areas were 72.7 per cent, 71.5 per cent, and 50.9 per cent, respectively, at the end of 2004. Overall, data on the Internet usage rate suggest that there is no significant difference in the access to Internet infrastructure by regions (i.e., between the Capital Region and non-Capital Region), but some difference exists between rural and urban areas. In addition, the difference between rural and urban areas may not be the result of the access to the Internet infrastructure, but the difference in the age distribution. In rural areas the proportion of population over age 65 years is much higher than in the urban areas. In 2000, the proportion of those belonging to the older ages in rural areas was 17.9 per cent, while that of cities was only 4.3 per cent (KRNIC 2003).

Even though the regional disparity in the usage rate is not significant, ".kr" domains are considerably concentrated in the Capital Region that includes Seoul city, Incheon city, and Gyeonggi province. According to the survey result of Korea Network Information Center (KRNIC 2005), in August 2005, Seoul was home to 55 per cent of the total number of .kr domains, and the Capital Region had 75.9 per cent of the total in Korea (Table 2.3). Even though the share of Seoul decreased slightly, the share of the Capital region increased slightly in those last two years. The higher concentration rate of .kr domains in the Capital Region compared with population or Internet users suggests that the concentration might be related to some other factors that attract IT-related firms in Seoul. The distribution of B2C e-marketplaces is also highly concentrated in the Capital Region. About 73 per cent of the total number of B2C e-marketplaces is concentrated in Seoul (Choi 2003).

Firms operating B2B e-marketplaces are more concentrated in Seoul and its surrounding areas than .kr domains. The Capital Region shares 79.5 per cent of the total firms operating B2B EC sites in Korea (Table 2.3). The Southeastern region, the second largest industrial zone in Korea, shares only 14.2 per cent of the national total. If we consider only the "public B2B e-marketplaces," in which many sellers and many buyers can access for transactions, the degree of concentration in the Capital Region is overwhelming. About 95 per cent of the firms operating public B2B e-marketplaces are located in the Capital Region (Choi 2003). The predominance of Seoul with about 84 per cent of the total public B2B

Table 2.3 Regional share of .kr domains and B2B EC sites

(Unit: per cent)

Region	Population (2000)	.kr Domain (2003)	.kr Domain (2005)	B2B EC sites (2003)
Capital R	46.3	74.9	75.9	79.5
- Seoul	21.5	56.2	55.0	63.0
- Incheon	5.4	3.4	3.4	3.0
- Gyeonggi	19.5	15.3	17.5	13.5
Middle R	10.1	4.7	4.9	2.6
- Daejeon	3.0	2.3	2.4	1.2
- Chungbuk	3.2	1.2	1.1	0.9
- Chungnam	3.9	1.2	1.4	0.5
Southwest R	11.3	4.6	4.0	3.7
- Gwangju	2.9	2.3	1.8	1.6
- Jeonbuk	4.1	1.3	1.3	1.9
- Jeonnam	4.3	1.0	0.9	0.2
Southeast R	27.9	14.2	13.2	14.2
- Busan	8.0	4.8	4.4	1.4
- Daegu	5.4	4.1	3.7	2.3
- Ulsan	2.2	0.9	0.8	2.6
- Gyeongbuk	5.8	2.0	1.9	3.5
- Gyeongnam	6.5	2.4	2.4	4.4
Gangwon	3.2	1.2	1.3	-
Jeju	1.1	0.5	0.5	-
Total	100.0	100.0	100.0	100.0

Note: The sum may not be 100.0 due to the rounding error.
Source: KRNIC, 2005 (recited from http://isis.nic.or.kr).

e-marketplaces might be related with the cluster of IT firms, many IT related spin-offs, and innovative entrepreneurs or talents in Seoul, especially in Gangnam area (Park 2004). IT related firms and advanced producer services are strongly concentrated in the Gangnam District within Seoul (Park and Choi 2005).

Even though the higher concentration ratio of firms operating the B2B EC sites in the Capital Region is a general pattern, there are some differences in the degree of the concentration by the nature of transaction through the B2B EC sites (Park 2004). The "open type" is more concentrated in Seoul than the "cooperative or closed type." In the open (bidding) type, any firm can participate as a buyer or as a seller, based on competitiveness via bidding, in the transaction through the B2B e-marketplaces. In the closed (cooperative) type, only permitted firms can participate in the transactions through e-marketplaces because the participating firms have had a long-term and stable contract with the operating firms of the B2B EC sites even before the launch of online transactions. The difference in the degree of the concentration in Seoul between the open type and the closed type suggests that the operating firms of the closed type of the B2B EC tend to be dispersed to industrial areas in order to facilitate local production networks. There is also a considerable difference between "buyer-centric" type and "seller-centric" type of the B2B EC. The overwhelming concentration of the seller type in the Capital Region is comparable to that of the B2C e-marketplaces. Compared to other types of B2B EC, the Southeastern Region shares a considerable higher percentage of the buyer-centric B2B EC sites. The relatively higher ratio of the buyer-centric types in the Southeastern Region seems to be related to the considerable concentration of manufacturing firms in this region, suggesting the network and cluster of firms in the industrial zones.

Spatial networks of private B2B electronic marketplaces: the case of SEC Electronics   The location pattern of firms that participate in the transaction through the private B2B EC will show actual network pattern of transactions. This section reviews the result of a recent analysis (Park 2004) of participating firms in the private B2B marketplaces operated by Samsung Electronics Company (SEC). SEC was selected because the electronics industry is the leading industry in private B2B EC in Korea, and SEC is regarded as a representative company operating private B2B marketplaces in the industry.

In 1997 SEC introduced Global Logistics Networks Systems (GLONETS) in order to rationalize the materials supplied from foreign countries. Internet purchasing systems through EDI was first introduced in 1997 in Korea by the development of the GLONETS. The introduction of the GLONETS system shortened the entry time to one day from one week. The Internet has contributed to integrate the SMART-NET and GLONETS into GLONETS in May 2000. Now GLONETS connects the suppliers throughout the world and developed an EC system that performs every stage of transactions. Only production materials such as parts and components are traded through the GLONETS. GLONETS operated by SEC is a prototype of private e-marketplaces run by a single large buyer. Although the web-based Internet system of GLONETS provides an opportunity to expand the ratio of transactions based on bidding, the opportunity is mainly for the suppliers that were already allowed to be potential trading partners by SEC. The firms attempting to be a new supplier of SEC should meet a certain level of qualification.

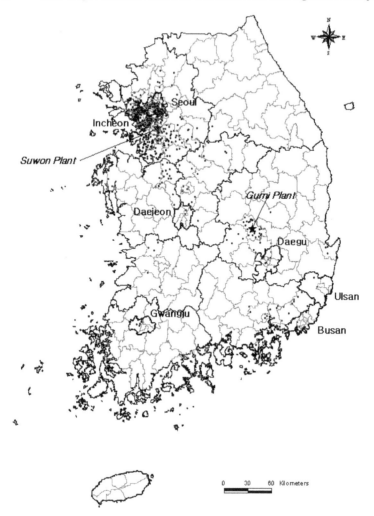

Figure 2.2   Cooperative suppliers of the Suwon Plant of SEC
Source: Park, 2004.

The higher concentration of cooperative suppliers to SEC in the Capital Region, which transacts through the private B2B marketplaces (GLONETS), is similar to the locational pattern of private B2B EC sites (Figures 2.2 and 2.3). About 80 per cent of cooperative suppliers to SEC's Suwon plant are concentrated in the Capital Region. It is surprising that the ratio of concentration in the Capital Region for the cooperative suppliers of the Gumi plant (about 82 per cent) is somewhat higher than the case of the Suwon plant. The higher concentration of the Suwon plant's suppliers in Gyeonggi represents the strong local transaction networks even in the GLONETS

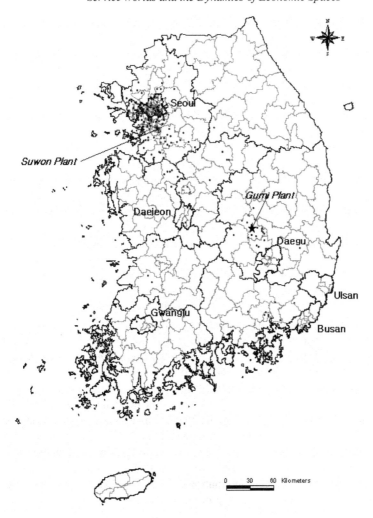

Figure 2.3   Cooperative suppliers of the Gumi Plant of SEC
Source: Park, 2004.

(Figure 2.2). However, in the case of the Gumi plant, the regional transaction network is not strong, but long-distance networks to Seoul and Gyeonggi are relatively strong, representing hierarchical pattern of networks (Figure 2.3). Areas near Suwon in Gyeonggi are in the top five in share of the cooperative suppliers to the Suwon plant at the county level. Gangnam and Seocho in Seoul and three counties in Gyeonggi are top five areas of share of the cooperative suppliers to the Gumi plant. In the international networks, Japan, the U.S., Taiwan, Hong Kong, and Singapore are the dominant countries in the share of cooperative suppliers through GLONETS. The foreign networks are mainly related to largest agglomeration areas of the electronics industry in the global economic spaces.

If we consider newly joined cooperative suppliers since 1999, the trends of strong local and regional networks for the Suwon plant and strong long-distance networks to Seoul city for the Gumi plant are more distinctive (Park 2004). The spatial network pattern of the newly joined suppliers strongly supports the intensification of local networks in the Capital region on the one hand and long-distance networks with industrial clusters or innovation centers in the non-Capital region on the other.

Overall, the location pattern of IT industries and e-commerce and spatial networks of SEC through GLONETS support the three aspects of the dynamics of economic spaces in the service worlds. First, the introduction of Internet has intensified the concentration of advanced services in Seoul and the production networks in the industrial zones of non-Capital region, suggesting the intensified spatial division of labor. Advanced services and IT services are overwhelmingly concentrated in Seoul (Park and Choi 2005) while production activites are fairly dispersed in the non-Capital region. The strong networks of SEC's Gumi plant with the Gangnam area of Seoul, where the advanced services and high tech spin-offs are concentrated, also suggests the intensified spatial division of labor in the Internet era.

Second, the clustering of the internet industry and advanced services is prominent in Seoul. It was clear that the overwhelming concentration of the .kr domain and firms operating EC was in Seoul in recent years. Since the late 1980s advanced services such as software, engineering, advertising, and design services have concentrated in the Gangnam area, the southern part of Han River in Seoul (Park and Nahm 1999). The concentration trend has culminated since the financial crises in 1997 with agglomeration of new start-ups in high tech and software sectors. Firms in Gangnam have strong localized networks of innovation even though they have the innovation networks in other various spatial scales such as regional, national, and international levels (Park 2005). Their economic behaviors are embedded culturally and socially in Kangnam. They have significant international networks of innovation, especially with the USA, but the collective learning processes within Gangnam area are more important for innovation than national and international networks. This suggests the evolution of a cluster of innovation and advanced services based on local networks and embeddedness.

Third, globalized networks of advanced services have been increased during the last decade. SEC emphasizes global sourcing through GLONET in recent years and has developed considerable networks with high tech areas in advanced countries. The role of foreign producer services firms in Seoul is significant (Lee 2001). The development of EC promotes long-distance networks including international networks on the one hand, but only to selected agglomeration centers of innovation centers, and intensification of local networks on the other.

The examples of the spatial networks of the B2B EC also suggest that transferability, networks and collaboration in the process of knowledge genesis, and hierarchical pattern of transaction are important for transactions in economic space.

Innovation Networks in Sunchang County

We may think that networks and embeddedness are important influences on spatial clustering and dispersal only in advanced economies, not applicable to peripheral areas even in the information society. However, these processes, based on the principles of spatial interaction in service worlds, are important in the evolution of new economic spaces in peripheral areas. The case of Sunchang-gun (county) of Jeonbuk province in Korea can illustrate this.

Sunchang is located in the mountainous and peripheral area of southwestern Korea. Sunchang is a part of the longevity area in Korea (Park, Jeong and Song 2005). The population of Sunchang has decreased continuously since the 1970s. Agriculture has been the key economic sector and the traditional agricultural products in this area were tobacco, red pepper, and diverse mountainous vegetables and fruits. Gochoojang (thick soypaste mixed with red pepper) has been a famous product of Sunchang because of its distinct taste that reflects the local fermentation process. Traditionally in Korea most households made their own Gochoojang. However, most households now buy Gochoojang, and the brand name "Sunchang Gochoojang" has become famous.

Since the late 1990s the Gochoojang production system has significantly changed. The Daesang company, among the largest food companies in Korea, established a branch plant in Sunchang to produce a standardized Gochoojang under the brand "Sunchang Gochoojan," maintaining traditional taste with high quality control. Daesang has advertised the quality of Sunchang Gochoojang and has continued R&D activities in order to produce high quality through automated mass production systems. The case of Daesang shows that production technology of the large company is linked to traditional local culture and resources. That is, the codified knowledge of Daesang is linked to locally embedded tacit knowledge and resources.

Simultaneously, Sunchang county created a real estate complex to gather traditional Sunchang Gochoojang makers together in one place. The county allowed skilled persons, licensed and with more than ten years experience making the traditional Gochoojang, to move their households into the complex. Fifty-four households moved to the complex to make their own specialized traditional Gochoojang. They have their own market to sell their products and also started to sell their products through the internet to consumers in large cities.

Manufacturing process and networks in making Gochoojang differ according to the size of firms. Traditional small and medium sized firms and family type traditional small manufacturers have considerably strong local or regional networks for raw materials, manpower, and information and knowledge for innovation. They produce Gochoojang based on the traditional way of manufacturing (with a long fermentation period; using local pepper and beans; non-standardized small scale of production). Large firms, however, produce standardized mass products with relatively short fermentation period and high rate of automation in the production process.

Traditional small and medium sized firms use mainly local input materials: 90 per cent of hot pepper and 62 per cent of beans from Sunchang county and

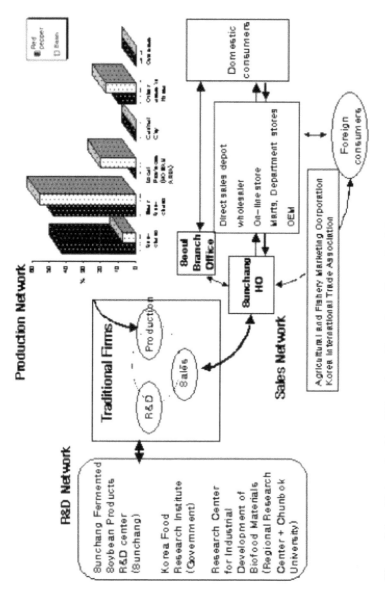

Figure 2.4   Network of traditional small- and medium-sized firms
Source: Based on interview survey and Lee (2005).

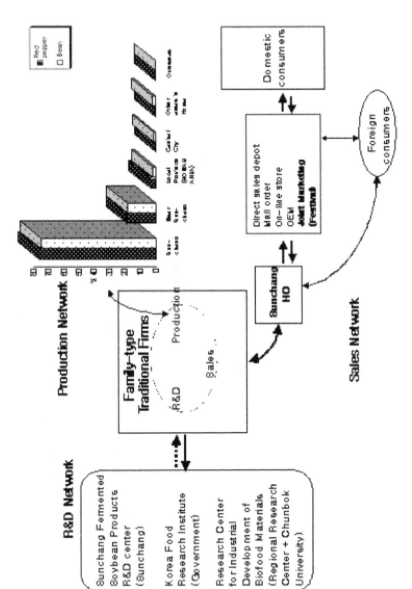

Figure 2.5   Networks of family-type traditional small businesses
Source: Based on interview survey and Lee(2005).

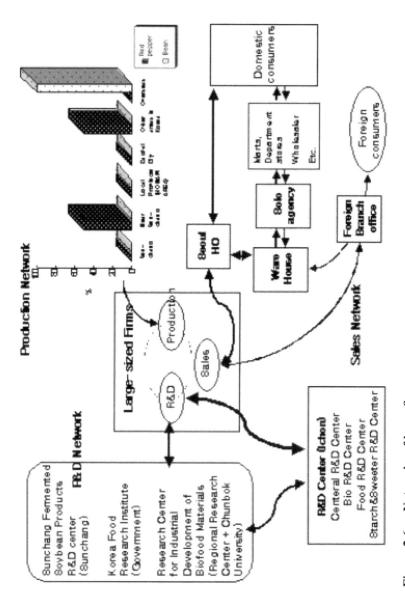

Figure 2.6   Networks of large firms
Source: Based on interview survey and Lee (2005).

adjacent counties (Figure 2.4). They have division of production, marketing and R&D within the firm. They have close cooperative relationships with the Research Center in Sunchang established by county government, Korean Food Development Research Institute, and Bio-Food research Center of Jeonbuk National University. Knowledge transfers continuously through the R&D networks between the firms and research institutions. They also have diverse marketing networks from national and international consumers. However, the most important customers are large department store and television shopping (Lee, 2005).

Family type small businesses have more strong local networks for raw materials: More than three quarters of hot pepper and bean are supplied from Sunchang and additional 18-19 per cent from adjacent counties (Figure 2.5), representing the family type manufacturers keeps traditional local culture in making Gochoojang with pure local raw materials. Activities of production, marketing and R&D are not clearly divided within business and have only weak networks with local research institutes and universities. Their marketing strategies are not much diversified and collaborative marketing with other small businesses is important. It should be noted that, however, they have strong collaborative networks within the local community: operating a joint web site for e-commerce, holding a festival for Sunchang Gochoojang, creating tourist experiences for Gochoojang making on site, and organizing an association for the conservation of traditional Gochoojang.

Large firms have strong networks with foreign bean suppliers. About half of the hot pepper is supplied form adjacent counties and another half from other regions in the nation (Figure 2.6). Large firms have a clear division of production, marketing, and R&D within firms and intensive networks of innovation. They largely depend on their own R&D centers located in the Capital region for the innovation. However, they have also quite strong networks with local and national research institutions (Figure 2.6). Marketing is mainly controlled by headquarters located in Seoul, with complex marketing networks for domestic and international consumers. However, they are embedded locally in the traditional culture of Gochoojang and the brand name of Sunchang.

Even though the strength of local networks of input materials and innovation is different for each type of firm, it is clear that collaborative networks for marketing and innovation are important for all the Gochoojang makers. Networks and collaboration are important in the production and service activities. For large firms, even though they have strong non-local networks for innovation and raw materials, they have strong networks with local community and culture for obtaining knowledge and information.

Sunchang-county also established a research lab in the complex to support quality control of the traditional Gochoojang. In the research lab, four researchers conduct quality control and research to develop new products using a local resource, raspberries. The idea of using raspberries instead of beans as an input material for fermentation originated from the fact that raspberry wine is good for health. To develop a new Gochoojang using raspberries, the researchers interacted with local traditional Gochoojang makers, Daesang company engineers, and an internet survey to gather information and knowledge. The researchers also interacted with faculty in Jeonbook National University located in Jeonjoo, the provincial capital, about 60 km distance from Sunchang. After synthesizing the information gathered, they conducted tests to analyze the effectiveness of the Raspberry-Gochoojang. Locally embedded resources and knowledge were networked with non-local institutions to secure the development of the new product. In this case, local networks and embeddedness were continuously linked to non-local institutions, suggesting the possibility of developing innovation and production systems even in such an underdeveloped peripheral area.

The Sunchang case suggests a new context for the organization of production systems and economic spaces in peripheral areas, developing new ideas based on traditional local knowledge, local innovation, and non-local networks. Local knowledge was supplemented with networks of advanced services research scientists and engineers outside the region. This organization of networks does not represent an actual cluster of advanced services within Sunchang county. However, the organization of networks made it possible to promote local innovation through a virtual innovation network.

The improvement of transferability and knowledge genesis made possible the development of virtual innovation networks in the peripheral area in Korea. Good highways improved the accessibility of Sunchang to major regional cities and to Seoul. In addition, high quality internet infrastructure contributed to the organization of periodic meetings of the high tech engineers and scientists. Traditional know-how about the Sunchang Gochoojang and Daesang's R&D activities contributed to the knowledge genesis in Sunchang. The four notions of interaction in the process of innovation combine to explicate the new organization of production systems and a virtual innovation cluster in Sunchang. The case of Sunchang suggests that the processes of forming new economic spaces are applicable not only to innovation centers of advanced countries, but also to developing areas in a knowledge-based information society.

Conclusion

The development of ICT has significantly increased the transferability of services in the knowledge-based economy, strengthening both globalization *and* localization forces. Knowledge-based advanced services have clustered in a few major cities, as global networks of service activities, has increased the availability of these services in far-flung locations. Firms' networks and embeddedness are important conduits for the clustering of service activities in major cities, but in the peripheral areas, these services can enter networks with local resources and culture, with significant economic impacts.

The cases of SEC's electronic commerce and innovation networks of Sunchang Gochoojang support the importance of the four notions of interaction in the internet era: transferability, knowledge-genesis for service products, network and collaboration, and hierarchy of control. The four notions dynamically shape economic spaces of both developed and less developed regions. The increase of transferability has contributed to the dispersion of economic activities from the core to the peripheral areas and long distance networks for products, services, and high quality manpower. However, the power of knowledge genesis, networks and collaboration in the major innovation centers or cities has promoted clustering of the knowledge-intensive activities within a few cities and then contributed to the intensification of the spatial division of labor. The intensification of clustering of advanced services in the major metropolitan areas has naturally resulted in the hierarchy of control in the economic spaces.

On the other hand, in peripheral areas, local culture and resources can be a source of knowledge genesis and can contribute to the innovation with improvement of transferability, collaboration among diverse actors through virtual innovation networks. The integration of the notions of interaction of the service worlds in the peripheral areas can promote the development of virtual innovation cluster. Accordingly, the dynamic integration of the four notions of interaction in the knowledge-based information society has resulted in the dynamic and controversial economic spaces in core and peripheral areas such as cluster and global networks, intensified spatial division of labor, and virtual innovation clusters.

The results of this study have several policy implications for the peripheral regions or developing areas. First, innovation networks at various spatial levels have become more important in both core regions and peripheral areas. The regional innovation systems approach can be applicable not only developed areas but also developing areas in a knowledge-based economy. Second, place-based marketing strategies using local culture and resources can be useful when the strategies are linked to innovation networks. Developing a local product's brand name with relation to the region can be a kind of place marketing strategy. Third, good information infrastructure in the peripheral areas is necessary. ICT facilitates economic development. However, it should be noted that advanced ICT infrastructure is a necessary, but not sufficient, precondition to stimulate growth. Fourth, training and retraining for local labor in the peripheral areas are needed to implement e-commerce marketing. Extending e-commerce for sales of local products in the peripheral areas is a useful strategy in the Internet era. Fifth, linking production process from raw materials to final consumption should be promoted in the peripheral areas. Even though the market of the peripheral area itself is small, the local products can be supplied to tourists in the local areas with promotion of tourism based on the experience of manufacturing and services of local products. Lastly, development of virtual innovation clusters and learning festivals are important in the peripheral areas. In the knowledge-based economy, peripheral areas have more disadvantages for economic development because innovation activity tends to cluster in the major innovation centers. Actual clustering of the innovation and advanced services in the peripheral areas is not easy. Because innovation activity can be applicable to the locally embedded culture and resources in the peripheral areas, virtual innovation clusters using virtual networks of innovation can be promoted in peripheral areas. In the construction of virtual innovation clusters, utilization of the diversity of locally embedded resources and culture is critical.

References

Beyers, W.B. (2002), 'Services and The New Economy: Elements Of A Research Agenda', *Journal of Economic Geography* 2.1, 1–30.

Bryson, J.R., Daniels, P.W. and Warf, B. (2004), *Service Worlds. People, Organizations, Technologies* (London: Routledge).

Castells, M. (2000), *The Rise of Network Society. The Information Age* Vol 1. 2nd edition. (Oxford: Blackwell).

Choi, J.S. (2003), *Public B2B Electronic Marketplaces: A Spatial Perspective.* Ph.D. Dissertation, Department of Geography, Seoul National University.

Christaller, W. (1966), *Central Places in Southern Germany* (translated by CW Baskins). (Englewood Cliffs, NJ: Prentice-Hall).

Clark, G.L. (1981), 'The Employment Relation and the Spatial Division of Employment', *Annals of the Association of American Geographers* 71, 412–24.

Clark, G.L. (2002), 'London in the European Financial Services Industry: Locational Advantage and Product Complementarities', *Journal of Economic Geography* 2, 433-453.

Clark, T. and Lloyd, R. (2000), *The City as an Entertainment Machine* (Chicago: University of Chicago Press).

Daniels, P.W., Ho, K.C. and Hutton, T.A. (eds) (2005), *Service Industries and Asia-Pacific Cities* (London: Routledge).

Hayter, R. (1997), *The Dynamics of Industrial Location* (Chichester: Wiley).

Gereffi, G. and Korzeniewicz, M. (eds) (1994), *Commodity Chains and Global Capitalism.* (Westport: Greenwood Press).

Glaeser, E.L., Kolko, J., and Saiz, A. (2001), 'Consumer city', *Journal of Economic Geography* 1, 27–50.

KNSO (Korean National Statistical Office) (2005), *Result of EC Statistical Survey* (Seoul: KNSO). (www.nso.go.kr).

KRNIC (Korea Network Information Center) (2003), http://isis.nic.or.kr.

Lee, K.J. (2005), *Analysis On The Production And Innovation Networks Of The Resource-Based Industries In Rural Areas In Korea – Two Case Studies On Hot Pepper Sauce Manufacturing In Sunchang And Wild-Berry Wine Manufacturing In Gochang.* MA thesis, Department of Geography, Seoul National University.

Lee, B.M. (2001), *Spatial Characteristics of Foreign Firms' corporate Networks in Korea: The Case of Business Service.* Ph.D. Dissertation, Department of Geography, Seoul National University.

Leinbach, T.R. and Brunn, S.D. (2001), *Worlds of E-commerce* (Chichester: Wiley).

Loesch, A. (1954), *The Economics of Location* (New Haven: Yale University Press).

Maskell, P. and Malmberg, A. (1999), 'Localized Learning and Industrial competitiveness', *Cambridge Journal of Economics* 23, 167–185.

Massey, D. (1984), *Spatial Divisions of Labor: Social Structure and the Geography of Production* (London: Macmillan).

Nonaka, I. and Takeuchi, H. (1995), *The Knowledge-Creating Company, How Japanese Companies Create the Dynamics of Innovation* (Oxford: Oxford University Press).

Park, S.O. (1993), 'Industrial Restructuring and the Spatial Division of Labor: The Case of the Seoul Metropolation Region, the Republic of Korea', *Environment and Planning A* 25.1, 81–93.

Park, S.O. (2003), 'Economic Spaces in the Pacific Rim: A Paradigm Shift and New Dynamics', *Papers in Regional Science.* 82: 2, 223–247.

Park, S.O. (2004), 'The Impact of Business-to-Business Electronic Commerce on the Dynamics of Metropolitan Spaces', *Urban Geography* 25.4, 289–314.

Park, S.O. (2005), 'Network, Embeddedness, and Cluster Processes of New Economic Spaces in Korea', in Le Heron, R. and Harrington, J. W. (eds) *New Economic Spaces: New Economic Geographies* (Aldershot, UK: Ashgate).

Park, S.O. and Choi, J. S. (2005), 'IT Service Industries and the Transformation of Seoul', in Daniels, P.W., Ho, K.C. and Hutton, T.A. (eds), *Service Industries and Asia-Pacific Cities* (Aldershot, UK: Ashgate).

Park, S.O., Jeong, E.J. and Song, K.U. (2005), 'Spatial Characteristics of Longevity Degree In Korea', *Journal of Korean Association of Regional Geographers* 11:2, 187–210.

Park, S.O. and Nahm, K.B. (1998), 'Spatial Structure and Inter-Firm Networks of Technical and Information Producer Services in Seoul, Korea', *Asia Pacific Viewpoints* 39: 2, 209–219.

Park, S.O, Song, K.U, and Jeong, E.J. (2005), 'Industrial and Innovation Networks of the Long-Live Area of Honam Region', *Journal of Korean Geographical Society* 40: 1, 78–95.

Park, S.O. and Wheeler, J.O. (1983), 'The Filtering Down Process in Georgia: The Third Stage of the Product Life Cycle', *The Professional Geographer* 35, 18–31.

Timmers, P. (2000), *Electronic Commerce: Strategies and Models for Business-to-Business Trading* (Chichester, UK: John Wiley).

Turban, E., King, D., Lee, J., Warkentin, M., and Chung, M. (2002), *Electronic Commerce: A Managerial Perspective* (New Jersey: Prentice Hall).

Ullman, E.L. (1957), *American Commodity Flow: A Geographical Interpretation of Rail and Water Traffic Based on Principles of Spatial Interaction* (Seattle: University of Washington Press).

Vernon, R. (1966), 'International Investment and International Trade in the Product Life Cycle', *Quarterly Journal of Economics* 80, 190–207.

Wheeler, J.O. and Mitchelson, R.L. (1989), 'Information Flows Among Major Metropolitan Areas in the United States', *Annals of the Association of American Geographers* 79: 4, 523–543.

Yeh, A.G.O. (2005), 'Producer Services and Industrial Linkages in the Hong Kong-Pearl River Delta Region', in Daniels, P.W., Ho, K.C., and Hutton, T.A. (eds), *Service Industries and Asia-Pacific Cities* (London and New York: Routledge).

Zook, M.A. (2002), 'Hubs, Nods and By-passed Places: A Typology of E-Commerce Regions in the United States', *Tijdschrift voor Economische en Sociale Geografie* 93: 5, 509–21.

Zook, M.A. (2005), *The Geography of the Internet Industry* (Oxford: Blackwell).

Chapter 3

Knowledge Intensive Services and R&D Diffusion: An Input-Output Approach

José A. Camacho and Mercedes Rodríguez

It has become increasingly evident that the tertiarisation process is continuing in all advanced economies. Service activities have increased their role in all dimensions of life: households consume more and more services, firms spend greater and greater amounts of money on services such as marketing and distribution, and the public sector provides an increasing variety of services. From a macroeconomic viewpoint, we notice that not only are service industries growing in their own right (and faster than the rest of the economy), but there also seems to be an increasing tendency for services to constitute the major part of the value added in most manufacturing industries (what is called *servicisation*) (Quinn et al. 1990). This tendency is particularly marked in the case of producer services, which are becoming the 'prime source of sustained high value added' (Gibbons et al. 1994, 12), and occurs to such an extent that, in many cases, it is extremely difficult to distinguish where the product ends and the service begins. This situation has given birth to terms such as *encapsulation* (Howells 2000) or *Rainbow Economy* (Coombs and Miles 2000; Boden and Miles 2000; Hauknes and Miles 1996). Coombs and Miles (2000) summarise the current situation as a process in which service firms are industrialising and manufacturing firms are becoming more service-like. Howells (2000b, 15) goes further and predicts the occurrence of a progressive *encapsulation*, understanding this as the phenomenon by which 'manufactured products are not offered to consumers in their own right, but rather in terms of more final demand'. He cites as examples the cases of the aerospace industry, which has substituted the offers of its products (engines) for services (hours of flight), or the computer industry, which offers computer services instead of computers themselves.

The question that inevitably arises is: What are the major reasons that explain this unceasing increase of services in production? Some authors (Miozzo et al. 2002; Andersen and Corley 2002; Howells 2000b) argue that the restructuring processes of many firms, increasing the volume of externalised service activities, is one of the major factors that explains the fast growth of services. However, in contrast to earlier periods, the functions that have been outsourced in recent years are highly skilled activities, requiring a well-trained labor force in order to be provided correctly, such as software consultancy, marketing or research and development. Furthermore, the type of relationship that ties manufacturing and services together is not one of subordination but rather of cooperation and collaboration. The argument for externalising with the aim of cost cutting is now obsolete, and most of the business services provided to firms are of strategic importance for the firm's performance in all domains (Porter 1990).

This increase of and change in the use of services reflect the development of the so-called Knowledge-based Economy. Knowledge has become the key asset in competition, and therefore those industries particularly related to the acquisition, diffusion and transmission of knowledge have assumed a preponderant role. The amount and diversity of knowledge that firms require has motivated much corporate restructuring, resulting in the outsourcing of many service functions. The clearest examples are found in the demand for new capacities and abilities caused by the information technologies revolution. Services are the main users of Information and Communication Technologies (ICT) (Moulaert et al. 1990; Litch and Moch 1999; Pilat 2000), and are also heavy investors in human capital (Illeris 2000; Miles and Tomlinson 2000; Camacho and Rodríguez 2005). These novel features convert them into key agents in order to introduce ICT and manage them, and in general to introduce and implement practically any new technology or technique.

Within the above context, we attempt, in this chapter, to carry out a macroeconomic approximation of the diffuser role that some service industries (namely the knowledge-intensive services, KIS) have assumed in six European countries: Denmark, Finland, France, Germany, Spain and United Kingdom. More concretely, by utilising the information supplied by R&D statistics and input-output tables, we try to establish a connection between two factors: (1) the increasing innovation efforts developed by KIS and (2) the rising interrelation among service industries and the rest of the production system. Our hypothesis is that the combination of greater R&D intensities and higher interaction manufacturing-services (reflected in high forward linkages) results in a considerable improvement in the position of KIS in terms of product-embodied R&D diffusion, and more generally, in a reinforcement of the role of KIS in the innovation domain. In other words, the impact of KIS innovation efforts is not restricted to the development of innovations by these industries alone, but extends to other branches as well.

The structure of the chapter is as follows: the first section below is devoted to presenting the theoretical framework on which the empirical work is based. The second section presents the methodology and the empirical analysis. In this part, after describing the data and methodology applied, we review the evolution of the share of the service sector in total business expenditures on research and development. Next, we differentiate among industries, so as to take into account the heterogeneity of the service sector. From this analysis we confirm the existence of a group of knowledge-intensive services, mainly research and development, computer and related activities, and, to a lesser extent, post and telecommunications. These industries will be our focus of attention when calculating the domestic embodied R&D flows generated by them in 1995 and 1999. The domestic embodied R&D flows per unit of value added show the diffuser role that they are increasingly playing, whereas the comparison of the total domestic embodied R&D flows confirms their rising importance within the production systems. The last section summarises the most relevant conclusions reached.

Services and Innovation: Creating and Diffusing Knowledge

Until very recently the belief has prevailed that services are laggards in terms of productivity, technology, and innovation. The vision of services based on what are called *peculiarities* (Miles 1993; 1996) such as the low levels of capital equipment or reduced technical skills, has led to a neglect of services' roles in innovation processes. Very few authors foresaw the key roles that some service industries have developed.

Machlup (1962) was one of the pioneers in pointing out the diffuser role of some services, although he failed to discern their innovative capacity. Some twenty years later, Gershuny and Miles (1983) recognised the potential impact of information technologies on services. Barras (1986) was the first to develop a theory of innovation in services, the *reverse product cycle* in which the learning processes associated with the introduction of information technologies lead to improvements in quality and finally, to new services. The Commission of the European Communities helped build a new vision of service innovation, recognising early (1990) that 'the development of business services is not an independent occurrence unconnected with changes in industry ... Service inputs have a specific role because they are an essential vector carrying the intelligence, information, technologies and innovations which are permanently needed ...'.[1]

Nevertheless, it was in the second half of the 1990s that the first exhaustive works on services and innovation came to light, claiming more attention for these activities, both in theoretical (Gadrey et al. 1995; Gallouj and Gallouj 1996; Gallouj 1997; 1998; Gallouj and Weinstein 1997; Miles et al. 1995; Miles 1996; Metcalfe and Miles 1997) and empirical terms (Sundbo 1998; Hipp et al. 1996).

Three main lines of analysis can be distinguished in these works (Coombs and Miles, 2000): assimilation, demarcation and synthesis. The assimilation view highlights the similarities between manufacturing and services and defends the application to services of the innovation theories and concepts designed for the manufacturing sector. In contrast, the demarcation perspective argues that services are very different from manufacturing, as is shown by the co-production that usually takes place when a service is provided. These profound differences make it inappropriate to employ traditional innovation theories to study innovation in services. Midway between these two opposite positions is the synthesis approach which maintains that neither of the two extremes is correct. It is evident that there are many differences between manufacturing and services, but it is also true that both sectors are more and more integrated, and that they influence each other. As a consequence, there is an urgent need for a more systemic vision to be adopted– one capable of integrating

---

1   In *Business Services in the European Community: Situation and Role.* Cited in Hauknes (1996, 135).

manufacturing and services. Sundbo and Gallouj (2000) and Gallouj (2002b) propose an almost identical taxonomy to the one elaborated by Coombs and Miles (2000), differentiating three perspectives: technologist, service-oriented and integrative. The technologist view is founded on theories about technological systems and takes as the core of its analysis the implementation of machinery and equipment, usually at the expense of neglecting non-technological innovations (which are precisely the most common innovations in service activities). The service-oriented approach centres on non-technological innovation, taking the pure services as its main domain, that is, those services in which intangibility and co-production are more clearly visible. Finally, the integrative view is based on the fact that the boundary between products and services is less and less clear-cut.

The increase in the number of studies on services since the mid-nineties along with the profound changes brought about by the Knowledge-based Economy have motivated a dramatic upsurge of interest in the innovation functions of services, especially of knowledge-intensive services (KIS). We can affirm that businesses are the protagonists of the innovation process in all its dimensions. They transform basic research into applied research, create new products and services and, in short, carry out the bulk of innovative activities. Given this fact, the special relationship that links KIS with their client businesses and their increasing knowledge and innovation intensity have combined within this knowledge-demanding context to make innovation performance co-dependent on the quality of the KIS provided.

But, what type of services can be called KIS? The term KIS, as a differentiated group of service activities, was introduced by Miles et al. in 1995. They were characterised as those service activities that share the following features (Boden and Miles 2000, 17): 'rely heavily upon professional knowledge, either are themselves primary sources of information and knowledge (reports, training consultancy, etc.) or use their knowledge to produce intermediary services for their clients' production processes (e.g., communication and computer services); and are of competitive importance and supplied primarily to businesses.'

It is precisely their special relationship to knowledge that differentiates KIS from other activities. They are innovative in their own right, but unlike the majority of highly innovative manufacturing activities, their major purpose is to facilitate innovation in other industries. Antonelli (2000, p.171) describes the way they operate as follows: they function as 'holders of proprietary "quasi-generic" knowledge, from interactions with customers and the scientific community, and operate as an interface between such knowledge and its tacit counterpart, located within the daily practices of the firm'. That is to say, they act as bridges for knowledge (Czarnitzki and Spielkamp 2000) or as Den Hertog and Bilderbeek (1998) put it, as a *second knowledge infrastructure*, even substituting for functions traditionally attributed to the public sector. Hauknes (2000) calls this diffuser role of KIS *induced innovation*, which indicates a new type of innovation that has to be added to the five Schumpeterian types of innovation.

Specifically, the KIS have three main ways of contributing to the knowledge base (Kox 2002): developing original innovations (technological and non-technological), diffusing knowledge (combining their own knowledge with the client firm's knowledge) and surpassing the problem of human capital indivisibilities (that is to say, facilitating the access to specialised knowledge by small firms). A highly educated workforce, in combination with strong efforts in innovation (not only R&D expenditures, but also training and acquisition of new technologies), allows these service industries to improve their client's knowledge bases, and at the same time, their own knowledge bases. For example, the study carried out by Windrum and Tomlinson (1999) for United Kingdom, Netherlands, Germany and Japan demonstrates that those countries where there are strong links between KIS and other industries obtain higher spillovers from service innovation.

In short, what the recent theories about service innovation put forward is that KIS are ever more knowledge-intensive and more innovative. Moreover, as a consequence of the inherent co-production that takes place in their provision, KIS act as 'bridges' for innovation and knowledge in their client firms, and in general as key agents within the innovation systems. Starting from this premise, the objective of the next section is to empirically evaluate the R&D diffuser role of KIS, by employing statistical information available on innovation efforts and linkages with other industries. R&D intensity is used to indicate innovation efforts, and input-output data are used to measure linkages with the rest of the production system.

Empirical Assessment of the Diffuser Role of Services in Europe

Data and Methodology

The data used in this analysis are drawn from three OECD databases: the Analytical Business Enterprise Research and Development database (ANBERD), the Structural Analysis database (STAN) and the OECD input-output database 2002. The ANBERD provides a consistent data set of business enterprise R&D. Its main advantage is that it overcomes the difficulties of international comparability and interruptions in the time series. Expenditure data from 1987 to 2001 by industry are available for 19 of the largest OECD R&D performing countries. The STAN database includes annual measures of output, labor inputs, investment and international trade. It is based on the International Standard Industrial Classification (ISIC) Rev. 3. and includes data for 29 OECD countries. Finally, although the OECD input-output database 2002 has not been officially released as yet, input-output tables for 14 countries – the G7, Australia, Netherlands, Denmark, Finland, Greece, Norway and Spain – have been available since 2002 on the OECD's On Line Information System (OLIS). The tables provided by the different countries have been converted to a 42 harmonised sector level, and refer to year 1995 in most of the cases. Nevertheless, as has been recognised (Ahmad 2002), there is some lack of coherence between the input-output tables and other economic and national accounts' aggregates, especially with the equivalent data (value-added and output) used in the STAN database, even after adjustments for FISIM (Financial Intermediation Services Indirectly Measured) have been made. For example, in 1997 the gross output of the US construction industry was $944 billion, on an IO basis. The equivalent STAN estimate is only $603 billion. This is the main reason why we have used input-output tables only to calculate the output multipliers, and we have preferred to use STAN data for calculating R&D intensities and total flows.

The selection of countries to examine has been made on the basis of the availability of R&D and input-output data. For example, in the case of Greece the input-output table for 1995 is available but Greece is not included in the ANBERD database. The final group of countries comprises the following countries: Denmark, Finland, France, Germany, Spain and the United Kingdom. The input-output tables refer to 1995, except in the cases of Denmark (1997) and the United Kingdom (1998). The scope of the analysis is restricted to the years 1995 and 1999, due to the limited availability of R&D data in services. The number of industries has been reduced to 27, of which 7 are services, in order to homogenise the ANBERD, STAN and input-output classifications.

The methodology employed (Rodríguez 2003) is a modified version of that introduced in the famous work of Papaconstantinou et al. (1996) and applied to service activities by Amable and Palombarini (1998). However, we adopt a supply-side vision: instead of assessing which industries incorporate more embodied R&D through the acquisition of intermediate inputs, we appraise which industries diffuse more embodied R&D through intermediate sales. Due to the limited information available on services, we restrict our analysis to domestic embodied R&D flows.

We define the R&D intensity in industry i as R&D expenditures (R_i) divided by value added (W_i):

$$r_i = \frac{R_i}{W_i} \qquad (i = 1, 2, ..., n) \qquad [1]$$

Given the output inverse matrix, introduced by Ghosh (1958), the equilibrium equation in the domestic supply model is:

$$X = W(I - B)^{-1} \qquad [2]$$

where X is the vector of domestic gross outputs, W is the vector of domestic primary inputs (value added) and $(I - B)^{-1}$ is the domestic output inverse (Ghosh) matrix. We can define the matrix of domestic embodied R&D diffusion, D, by introducing the diagonalised matrix of R&D intensities in equation [2] as follows:

$$D = W\hat{r}(I - B)^{-1} \qquad [3]$$

where r-hat ($\hat{r}$) indicates a diagonal matrix whose elements are those of vector r.

Equation [3] relates domestic embodied R&D diffusion to the value added components (compensation of the employees and gross operating surplus). Thus domestic embodied R&D diffusion per unit of value added of industry i, UD_i, can be defined as the sum of the ith row of matrix $\hat{r}(I - B)^{-1}$:

$$UD_i = \sum_{j=1}^{n} r_i q_{ij} \qquad (i = 1, 2, ..., n) \qquad [4]$$

where q_{ij} are the elements of the Ghosh inverse. Since the ith row of the Ghosh inverse measures the impact on domestic production when the utilisation of primary inputs (valued added) of the ith industry varies by one unit, equation [4] provides the amount of domestic embodied R&D diffused per unit of value added of industry i. We can obtain the total domestic embodied R&D flows diffused through the intermediate purchases, TD, by pre-multiplying equation [4] by the value added:

$$TD_i = w_i \sum_{j=1}^{n} r_i q_{ij} \qquad [5]$$

where w_i is the value added of industry i.

The calculation of the flows per unit of valued added is a way of approximating the relevance of services as transmitters of their own R&D efforts. In a complementary way, the computation of the total flows allows us to evaluate the relative importance of service industries within the global diffusion process that takes place in the innovation systems.

Evolution of R&D Intensities in Services

Numerous problems emerge when we try to measure the innovation efforts in the service domain using the statistics available (Kleinknecht 1996; 2000). The well-known work of Allison Young (1996), where several OECD databases are combined in order to analyse the evolution of R&D expenditures in service industries, demonstrates that many of the differences in R&D intensities growth rates are caused by discrepancies in the statistical coverage. Moreover, much innovative expenditure in services are concentrated in areas not directly related to R&D, where the availability of data is quite scarce, specifically in the investment in human capital and the acquisition of information technologies and software (Pilat 2000; Sirilli and Evangelista 1998; Marklund 2000).

In spite of these drawbacks, if we observe the evolution of the average annual growth rates for R&D performed by the business sector during the second half of the nineties (Table 3.1), the growth has been higher in services than in manufacturing for all the European countries examined except the United Kingdom. On average, the growth of BERD in services in the European Union is three times the rate

Table 3.1 Percentage of BERD performed in the service sector, 1995–2000

	1995	2000*	Growth, services	Growth, manufacturing
Denmark	31.27	35.17	14.1	9.3
Finland	9.01	11.95	24.8	17.1
France	7.15	9.06	9.2	2.1
Germany	3.53	7.78	23.8	4.9
Spain	12.87	35.34	37.5	7.3
United Kingdom	17.99	16.55	3.0	5.2
EU	9.40	12.97	15.2	5.3

* Denmark, France, EU: 1999.

Source: ANBERD.

of growth experienced by manufacturing. Spain and Germany showed the most remarkable difference in R&D growth rates for the two sectors. Spanish growth rates of BERD during the period 1995–2000 were 37.5 per cent in services and 7.3 per cent in manufacturing. In Germany, the growth rates were 23.8 per cent in services and 4.9 per cent in manufacturing.

The OECD (2000) explains this rapid growth as a combination of three major causes:

- The advances in the elaboration of service statistics.
- The great number of research activities that services are carrying out in comparison with the past, either aimed at developing new services or improving the existent ones.
- The outsourcing or externalisation of research and development to service firms, by both firms and governments.

Nevertheless, despite these increases, services still account for a much smaller share of R&D than of GDP or employment and there are significant differences among countries. For instance, whereas in Denmark or Spain the service sector performs more than 35 per cent of BERD, traditionally manufacturing-oriented countries such as Germany or France show percentages below the European average.

Entering into more detail, we can examine the evolution of the R&D intensities in the different service industries, measured here as the ratio between business expenditures on research and development and value added.

Table 3.2 Evolution of R&D intensities in service industries, 1995–1999

	Denmark	Finland	France	Germany	Spain	United Kingdom
1999						
Wholesale and retail	0.93	0.11	n.a.	0.07	0.03	n.a.
Transport and storage	n.a.	0.18	1.28	0.65	0.01	n.a.
Post and telecom	2.14	4.89	n.a.	n.a.	1.63	2.34
Finance, ins.	0.42	n.a.	n.a.	0.01	n.a.	n.a.
Computer	10.29	4.60	2.00	3.00	1.96	3.72
R&D	36.35	n.a.	n.a.	14.25	11.62	12.05
Other services	0.54	n.a.	n.a.	0.11	0.15	0.13
Manufacturing	5.96	8.60	7.07	7.48	2.10	5.90
1995						
Wholesale and retail	0.45	n.a.	n.a.	0.02	0.005	n.a.
Transport and storage	n.a.	0.10	1.18	0.34	0.03	n.a.
Post and telecom	1.42	2.45	n.a.	n.a.	0.58	2.26
Finance, ins.	n.a.	n.a.	n.a.	0.03	n.a.	n.a.
Computer	6.78	6.78	2.97	0.63	2.11	7.42
R&D	14.77	n.a.	n.a.	3.90	2.31	8.27
Other services	0.96	n.a.	n.a.	0.11	0.17	0.25
Manufacturing	4.69	5.64	7.18	6.68	1.70	5.12

Source: Authors' calculations.

Table 3.3 Domestic product-embodied R&D diffused per hundred units of value added, 1995–1999

	1999						1995					
	DK	FI	FR	GE	SP	UK	DK	FI	FR	GE	SP	UK
Food, beverages and tobacco	1.89	4.69	1.47	0.83	0.90	1.67	2.05	3.20	1.38	0.87	0.71	1.48
Textiles, leather and footwear	1.02	2.95	1.48	2.88	1.02	0.54	0.23	1.81	1.47	2.01	0.53	0.40
Wood and products of wood and cork	1.02	1.98	0.99	0.73	0.23	n.a.	0.60	1.77	0.85	0.79	0.26	n.a.
Pulp, paper, printing and publishing	n.a.	2.57	0.77	0.66	0.90	n.a.	n.a.	2.12	0.71	0.75	0.43	n.a.
Coke, refined petroleum products	n.a	14.65	9.66	3.77	2.96	15.51	n.a.	9.98	7.50	5.73	2.52	15.94
Chemicals	32.16	22.64	23.34	25.36	8.58	34.37	22.79	21.23	21.87	20.66	6.93	26.31
Rubber and plastics products	7.51	10.23	9.95	5.95	3.34	1.99	3.09	8.03	7.67	4.23	2.74	1.93
Other non-metallic mineral products	2.45	4.76	4.56	4.48	1.31	2.66	1.36	8.72	4.00	3.41	0.95	2.35
Basic metals	4.86	7.43	6.84	3.25	2.40	3.03	2.32	5.59	7.69	3.33	1.64	2.40
Fabricated metal products	1.30	5.00	1.77	2.83	1.53	1.12	0.73	4.09	2.37	2.27	1.05	1.88
Machinery and equipment, n.e.c.	9.49	12.97	6.38	8.50	5.30	6.75	9.96	9.06	7.21	8.11	4.54	6.34
Office, accounting machinery	20.45	27.14	19.73	30.37	8.73	3.99	23.07	26.15	21.23	47.87	4.35	5.03
Electrical machinery and apparatus	13.21	22.02	11.34	6.37	6.01	9.26	5.61	17.35	11.68	13.02	3.88	14.54
Radio, television and com. equipment	17.79	43.20	43.63	53.28	22.96	15.40	25.46	46.40	44.26	51.89	20.48	14.27

	Medical, prec. & opt.	Motor vehicles	Building & equip.	Manufacturing n.e.c.	Electricity, gas, water	Construction	Wholesale & retail	Transport & storage	Post & telecom	Finance, insurance	Computer & related	R&D	Other business
Medical, precision and optical instrum.	10.32	11.99	24.89	n.a.	2.53	0.04	n.a.	n.a.	5.46	n.a.	18.25	18.18	0.51
Motor vehicles, trailers	5.77	3.22	18.01	0.74	1.99	0.03	0.01	0.06	1.16	n.a.	3.12	4.96	0.29
Building and equipment	20.80	16.90	62.03	1.78	0.63	0.09	0.03	0.62	n.a.	0.08	1.26	5.04	0.22
Manufacturing n.e.c.; recycling	45.79	21.35	47.68	1.48	2.43	0.28	n.a.	2.36	n.a.	n.a.	7.11	n.a.	n.a.
Electricity, gas and water supply	18.30	6.22	6.12	2.82	1.81	0.50	n.a.	0.21	5.44	n.a.	6.78	n.a.	n.a.
Construction	15.47	n.a.	2.71	n.a.	0.32	0.08	0.68	n.a.	3.07	n.a.	13.33	23.92	1.66
Wholesale and retail trade; repairs	13.43	14.77	26.07	n.a.	2.02	0.17	n.a.	n.a.	5.65	n.a.	9.15	26.48	0.26
Transport and storage	4.88	3.42	21.45	1.30	0.75	0.02	0.05	0.01	3.25	n.a.	2.89	24.89	0.26
Post and telecommunications	18.19	24.19	37.05	1.92	0.54	0.12	0.11	1.20	n.a.	0.03	5.98	18.43	0.22
Finance, insurance	29.57	17.94	37.22	2.60	3.22	0.43	n.a.	2.56	n.a.	n.a	4.78	n.a.	n.a.
Computer and related activities	11.55	4.43	3.57	3.08	4.10	0.70	n.a.	n.a.	10.87	n.a.	4.60	n.a.	n.a.
Research and development	18.18	n.a.	16.33	n.a.	0.39	0.09	1.38	n.a.	4.61	0.90	20.24	58.84	0.94
Other business activities													

Source: Authors' calculations.

Table 3.4 Domestic product-embodied R&D diffused, 1995–1999 (millions of national currency and pre-EMU euro)

	1999						1995					
	DK	FI	FR	GE	SP	UK	DK	FI	FR	GE	SP	UK
Food, beverages and tobacco	540.9	90.8	458.8	315.2	131.4	337.9	570.3	67.1	408.5	314.5	92.5	269.5
Textiles, leather and footwear	52.9	17.5	155.0	283.2	76.7	38.1	12.6	10.3	169.3	217.2	33.0	31.3
Wood and products of wood and cork	50.5	25.0	36.2	59.9	6.0	n.a.	25.4	19.8	29.4	69.4	5.2	n.a.
Pulp, paper, printing and publishing	n.a.	149.2	144.6	215.7	70.8	n.a.	n.a.	120.2	126.9	214.7	28.1	n.a.
Coke, refined petroleum products	n.a.	25.6	495.7	95.1	58.5	413.4	n.a.	25.4	389.6	131.9	52.4	466.0
Chemicals	6172.1	371.1	5731.3	9783.9	723.0	5211.6	3327.5	295.8	4825.2	8328.1	505.2	4027.7
Rubber and plastics products	648.0	95.6	1064.0	1220.6	153.8	157.0	208.5	51.8	701.4	780.0	96.5	130.8
Other non-metallic mineral products	215.8	39.3	490.1	746.1	100.5	132.2	97.7	51.3	398.7	613.1	58.0	121.0
Basic metals	167.1	72.6	515.6	511.2	93.6	128.2	82.7	71.6	618.0	532.5	65.0	134.3
Fabricated metal products	194.6	79.8	392.5	1061.5	122.0	138.5	96.3	49.1	461.3	792.7	62.8	197.9
Machinery and equipment, n.e.c.	2289.1	357.1	1141.9	4991.6	334.2	857.3	2391.5	224.9	1110.6	4491.3	212.2	778.5
Office, accounting machinery	206.8	10.3	476.8	1129.7	65.7	122.4	143.7	35.0	600.7	1876.4	33.7	165.3
Electrical machinery and apparatus	823.2	186.5	1121.4	1908.2	218.8	544.4	305.4	127.0	972.0	3629.9	105.9	753.4
Radio, television and com. equipment	937.3	1981.9	3214.6	5243.0	295.3	1040.8	915.3	675.6	2518.0	3897.0	250.1	722.7

Medical, precision and optical instrum.	1225.4	73.4	2241.0	2566.3	62.3	681.0	791.2	75.6	3073.3	2491.4	52.6	436.3
Motor vehicles, trailers	n.a.	15.2	3410.5	12 351.2	324.0	1336.5	n.a.	15.6	2941.3	7505.0	225.7	1002.4
Building and equipment	418.6	21.2	3210.7	3286.1	464.5	1790.2	116.4	38.8	3189.3	3244.0	293.0	1171.5
Manufacturing n.e.c.; recycling	n.a.	18.5	220.4	229.6	57.0	n.a.	n.a.	14.9	132.3	209.5	23.5	n.a.
Electricity, gas and water supply	93.5	90.1	865.5	213.8	96.5	321.9	69.4	42.2	630.8	228.0	239.2	394.8
Construction	52.8	42.5	232.5	121.8	9.8	69.6	33.4	19.3	161.1	102.7	8.9	13.6
Wholesale and retail trade; repairs	1843.2	19.5	0.0	206.5	28.2	n.a.	799.5	n.a.	n.a.	52.4	3.4	n.a.
Transport and storage	n.a.	27.9	1349.9	756.4	3.7	n.a.	n.a.	13.1	1012.4	344.2	14.5	n.a.
Post and telecommunications	1133.3	345.4	n.a.	n.a	449.4	1363.4	559.1	101.4	n.a.	n.a.	118.4	999.0
Finance, insurance	478.6	n.a.	n.a.	31.7	n.a	n.a.	n.a.	n.a.	n.a.	67.0	n.a.	n.a.
Computer and related activities	2973.6	74.2	1111.2	1724.1	135.0	1754.2	1591.1	54.0	1014.0	223.1	78.9	1660.7
Research and development	1952.3	n.a.	n.a.	1113.1	43.1	984.5	518.0	n.a.	n.a.	235.9	6.6	542.8
Other business activities	1638.6	n.a.	n.a.	917.6	179.2	403.9	2434.4	n.a.	n.a.	797.9	151.8	545.3

Source: Authors' calculations.

Table 3.2 shows the R&D intensity of service industries in 1995 and 1999. As can be seen, R&D intensity is far below that of manufacturing except in the industry of research and development and, in Denmark, in the industry of computer and related activities. From 1995 to 1999 most service industries have experienced high average annual growth of their R&D intensities. Post and telecommunications and research and development showed the highest growth rates. The case of computer and related activities draws our attention because of the contrasting pattern presented in the countries examined. Thus, whereas in Demark and Germany its R&D intensity has risen dramatically, it has declined in the other four countries.

In any case, we observe that the three industries mentioned (post and telecommunications, computer and related activities, and research and development) are the most R&D intensive or knowledge-intensive. Although a further breakdown in categorising would be more accurate, we will call these three industries KIS. The analysis of the domestic embodied R&D diffused by KIS will indicate whether these activities are also a source of knowledge (in this case of R&D) for other industries, as the theory presented in Section 2 argues.

Services and Embodied R&D Flows

The application of the methodology described above allows us to compare the evolution of the "diffuser" role of services during the period 1995–1999. Drawing on the hypothesis of the stability of the input-output multipliers (Bon, 2001), we calculate the domestic embodied R&D flows diffused for the years 1995 and 1999. The results are shown in Tables 3.3 and 3.4.

Table 3.3 reports the domestic embodied R&D diffused per hundred units of value added by industries in 1995 and 1999.

There has been a general improvement in the position of KIS in the European countries examined, particularly marked in the branch of research and development which is the leading diffuser industry in Denmark and Spain and occupies the second position in the United Kingdom. Post and telecommunications and computer and related activities have relevant positions too, along with highly innovative manufacturing industries such as medical, optical and precision instruments or electrical machinery.

Among countries, the strong increases in service R&D expenditures in Germany and Spain (Table 3.1) are clearly reflect in their diffuser activity. For instance, the three top industries in terms of average annual growth rates in Germany and Spain were service industries: computer and related activities, wholesale and retail trade, and research and development in Germany and wholesale and retail trade, research and development and post and telecommunications in Spain. Each exhibited annual average growth rates higher than 30 per cent.

Whereas in 1995 KIS played a remarkable role only in Denmark and the United Kingdom, the faster growth rate of KIS R&D intensities experienced in the remaining countries has caused the role of KIS to be quite similar across all Europe. Thus, the research and development industry, along with highly innovative manufacturing industries, carries out the most outstanding diffuser role in all countries, and the two other KIS industries, post and telecommunications and computer and related activities take on upper-intermediate roles.

The combination of services' growth in value added and R&D intensity has increased their diffuser role. This is indicated by the changes in the rank of diffuser industries in terms of total domestic embodied R&D flows (Table 3.4).

As might be expected, the upsurge in the unitary flows shown above, in addition to the growth of tertiary activities, has resulted in a considerable rise of the total domestic embodied R&D flows diffused by services, and more specifically by KIS. Thus, the three service industries grouped as KIS have improved their positions as diffusers in all of the countries, with the sole exception of the computer and related activities industry, which has dropped very slightly in the United Kingdom and France from the second to the third position and from the seventh to the ninth, respectively. We also observe that post and telecommunications and computer and related activities industries have moved to higher positions than those they occupied in terms of unitary flows, thanks to their large share in value added.

The most relevant point to highlight is the key role that KIS are exercising in terms of total embodied R&D diffusion. In the majority of cases, they are among the top-ten diffuser activities. There is, moreover, a high diffuser potential in some type of KIS that still demonstrate a low participation in the production system (the most striking case being the industry of research and development in Spain which descends from the leading position in terms of unitary flows to the twenty-second position in terms of total flows). Given their great diffuser power, their growth should translate into an increase in the embodied R&D acquired by their client industries and finally into better innovative performance.

Discussion

This chapter has tried to shed some light on the R&D diffuser role that some service industries, the knowledge intensive services (KIS), are playing within production systems.

The recent development of statistics on innovation that include service activities (including R&D statistics and innovation surveys) has allowed us to confirm empirically that services are not mere users of innovations, but also relevant creators and diffusers of knowledge. Moreover, despite the existence of a considerable heterogeneity among service industries and their tendency towards non-technological innovation, there are some service branches that present innovation patterns very similar to highly innovative manufacturing activities. This is the case of those services called knowledge intensive (KIS).

The theoretical consequence is that the polarisation of *assimilation* versus *demarcation* in the scope of service innovation is giving way to an intermediate position, called *synthesis* or *integration*, which highlights the blurring of boundaries between manufacturing and services (Drejer, 2004).

In consonance with these arguments, the empirical analysis carried out here has demonstrated that KIS (in this chapter research and development, computer and related activities and post and telecommunications) can greatly contribute to the innovation performance of other industries, by supplying product-embodied R&D.

KIS are ever more involved in the development of R&D activities, as shown by the great growth rates of their expenditures on research and development during the second half of the nineties. Thus, for example, in the European Union the growth rate of R&D in services was three times the R&D growth rate in manufacturing. The increase of R&D intensities in KIS industries specifically has been superior to those shown in manufacturing industries.

In addition, the orientation of KIS towards intermediate demand rather than final demand (or according to input-output terminology, their high forward linkages) causes them to be key diffusers of their increased innovation. KIS industries, and especially post and telecommunications and computer and related activities, hold leading positions in terms of total domestic product-embodied R&D flows, being among the top-ten diffuser industries in all of the countries analysed. In addition to this key role, there are potential benefits associated with the growth of KIS industries that are leaders in R&D diffusion per unit of value added, but currently show modest participation in production/value added. The clearest example of this can be found in the branch of research and development. Since it occupies a pre-eminent position in terms of unitary flows, growth in this branch will result in a considerable rise in total product-embodied R&D flows and, in general, in an improvement of innovation performance. Although the variable employed for calculating the diffusion flows has been the R&D expenditures, this merely opens the road for further analyses, because as has been stated before, most innovation expenditures in services are devoted to other areas, especially towards human capital formation. Although our analysis has been restricted to the domestic domain, the internationalisation of technology and knowledge is an important feature of our Knowledge-based economies, and should be the subject of future research.

Moving beyond R&D, we can affirm that there is a general trend for KIS to assume an increasing role in the diffusion of knowledge, particularly of a tacit type, to various agents (individuals, firms and industries) in all advanced economies. The co-production that has commonly been pointed out as one of the main characteristics of services is at the root of this diffuser function: given the close relationship established between provider and consumer, KIS can transmit and create knowledge. Therefore, more attention will have to be paid to these service activities to support their significant roles in innovation.

References

Ahmad, Nadim, 'The OECD input-output database', paper presented at the 14th International Conference on Input-Output Techniques, Montreal, 10–15 October 2002.

Amable, Bruno and Palombarini, Stefano (1998), 'Technical change and incorporated R&D in the services sector', *Research Policy* 27, 655–675.

Andersen, Birgitte and Corley, Marva, 'The Theoretical, Conceptual and Empirical Impact of The Service Economy: A Critical Review', paper presented at the 9th International Joseph Schumpeter Society Conference, Gainesville, 28–30 March 2002.

Antonelli, Cristiano (2000), 'New Information Technology and Localized Technological Change in the Knowledge-Based Economy', in Boden, Mark and Miles, Ian (eds), *Services and the Knowledge-based Economy*, London: Continuum, pp. 170–191.

Barras, Richard (1986), 'Towards a Theory of Innovation in Services', *Research Policy* 15, 161–173.

Boden, Mark and Miles, Ian (2000), *Services and the Knowledge-based Economy*, London: Continuum.

Bon, Ranko (2001), 'Comparative Stability Analysis of Demand-Side and Supply-Side Input-Output Models: Towards an Index of economics "Maturity"', in Lahr, Michael and Dietzenbacher, Erik (eds), *Input-Output Analysis: Frontiers and Extensions*, London: Palgrave, pp. 338–347.

Camacho, José A. and Rodríguez, Mercedes (2005), 'Human capital in the service sector and regional development: a comparison for the Spanish regions', *The Service Industries Journal* 25, 563–577.

Coombs, Rod and Miles, Ian (2000), 'Innovation, measurement and services: the new problematique', in Metcalfe, John S. and Miles, Ian (eds), *Innovation Systems in the Services Economy: Measurement and Case Study Analysis*, Boston: Kluwer Academic Publishers, pp. 85–103.

Czarnitzki, Dirk and Spielkamp, Alfred (2000), 'Business services in Germany: bridges for innovation', *ZEW Discussion paper* 52, Mannheim: ZEW.

Den Hertog, Pim and Bilderbeek, Rob (1998), 'The New Knowledge Infrastructure: The Role of Technology-Based Knowledge Intensive Business Services in National Innovation Systems', in Boden, Mark and Miles, Ian (eds), *Services and the Knowledge-based Economy*, London: Continuum, pp. 222–246.

Drejer, Ina (2004), 'Identifying innovation in surveys of services: a Schumpeterian perspective', *Research Policy* 33, 551–562.

Gadrey, Jean, Gallouj, Faïz and Weinstein, Oliver (1995), 'New modes of innovation: how services benefit industry', *International Journal of Service Industry Management* 6, 4–16.

Gallouj, Camal and Gallouj, Faïz (1996), *L'innovation dans les services*, Paris: Editions Economica.

Gallouj, Faïz (1997), 'Towards a neo-Schumpeterian theory of innovation in services', *Science and Public Policy* 24, 405–420.

Gallouj, Faïz (2000), 'Beyond Technological Innovation: Trajectories and Varieties of Services Innovation', in Boden, Mark and Miles, Ian (eds), *Services and the Knowledge-based Economy*, London: Continuum, pp. 129–145.

Gallouj, Faïz (2002), *Innovation in the Service Economy. The New Wealth of the*

Nations, Cheltenham: Edward Elgar.

Gallouj, Faïz and Weinstein, Oliver (1997), 'Innovation in services', *Research Policy* 26, 537–556.

Gershuny, Jonathan and Miles, Ian (1983), *The New Service Economy: The Transformation of Employment in Industrial Societies*, London: Pinter.

Ghosh, Ambica (1958), 'Input-Output Approach to an Allocative System', *Economica* 25, 58–64.

Gibbons, Michael, Limoges, Camille, Nowotny, Helga, Schwartzman, Simon, Scott, Peter and Trow, Martin (1994), *The New Production of Knowledge*, London: Sage.

Hauknes, Johan (2000), 'Dynamic Innovation Systems: What Is the Role of Services?', in Boden, Mark and Miles, Ian (eds), *Services and the Knowledge-based Economy*, London: Continuum, pp. 38–63.

Hauknes, Johan and Miles, Ian (1996), 'Services in European Innovation Systems – A review of issues', *STEP Report* 6, Oslo: STEP.

Hipp, Christianne, Kukuk, Martin, Licht, George and Muent, Gunnar (1996), 'Innovation in services. Results of an innovation survey in the German service industries', paper presented at the OECD Conference on the new S&T indicators for the knowledge-based economy, Paris, 19–21 June 1996.

Howells, Jeremy (2000), 'Innovation & Services: New conceptual frameworks', *CRIC Discussion Paper* 38, Manchester: The University of Manchester and UMIST.

Illeris, Sven (2000), 'Skills in Services: A Study in Denmark', *Service Development, Internationalisation and Competences Working Paper* 12, Roskilde: Center for Service-Studier.

Kleinknecht, Alfred (1996), *Determinants of Innovation*, London: Macmillan Press.

Kleinknecht, Alfred (2000), 'Indicators of Manufacturing and Service Innovation: Their Strengths and Weaknesses', in Metcalfe, John S. and Miles, Ian (eds), *Innovation Systems in the Services Economy: Measurement and Case Study Analysis*, Boston: Kluwer Academic Publishers, pp. 169–186.

Kox, Henk, 'Innovation and Knowledge Dissemination Keep Baumol Disease Away: The Case of Business Services', paper presented at the 12th International RESER Conference, Manchester, 26–27 September 2002.

Litch, George and Moch, Dietmar (1999), 'Innovation and Information Technology in Services', *Canadian Journal of Economics* 32, 363–383.

Marklund, Goran (2000), 'Indicators of Innovation Activities in Services', in Boden, Mark and Miles, Ian (eds), *Services and the Knowledge-based Economy*, London: Continuum, pp. 86–108.

Miles, Ian (1993), 'Services in the new industrial economy', *Futures* 25, 653–672.

Miles, Ian (1996), *Innovation in Services: Services in Innovation*, Manchester: Manchester Statistical Society.

Miles, Ian and Tomlinson, Mark (2000), 'Intangible Assets and Service Sectors: The Challenges of Service Industries', in Buigues, Pierre, Jacquemin, Alex and Marchipont, Jean Francois (eds) *Competitiveness and the Value of Intangible Assets*, Cheltenham: Edward Elgar, pp. 154–186.

Miles, Ian, Kastrinos, Nikos, Bilderbeek, Rob, den Hertog, Pim, Flanagan, Kieron, Huntink, Willem and Bouman, Mark (1995), *Knowledge-intensive business services: their role as users, carriers and sources of innovation*, report to the EC DG XIII Sprint EIMS Programme, Luxembourg.

Miozzo, Marcela, Ramirez, Paulina and Grimshaw, Damian 'High Tech Business Services and Innovation: New Organizational Forms and the Challenges of Global Economic Change', paper presented at the 12th International RESER Conference, Manchester, 26–27 September 2002.

Moulaert, Frank, Martinelli, Flavia and Djellal, Faridah (1990), *The Role of Information Technology Consultancy in the Transfer of Information Technology to Production and Service Organizations*, Amsterdam: Nederlandse Organisatie voor Technologisch Aspectenonderzoek.

OECD (2000), *Science, Technology and Industry Outlook 2000*, Paris: OECD.

Papaconstantinou, George, Sakurai, Norihisa and Ioannidis, Evangelos (1996) 'Embodied technology diffusion: an empirical analysis for 10 OECD countries', *STI Working Paper* 1, Paris: OECD.

Pilat, Dirk (2000), 'Innovation and Productivity in Services: State of the Art', in OECD *Innovation and Productivity in Services*, Paris: OECD.

Porter, Michael (1990), *The competitive advantage of nations*, London: Macmillan.

Quinn, James Brian, Doorley, Thomas L. and Paquette Penny C. (1990), 'Technology in Services: Rethinking Strategic Focus', *Sloan Management Review* 11, 79–88.

Rodríguez, Mercedes (2003) 'Services and Innovation: Towards the Knowledge Economy', unpublished PhD thesis, University of Granada.

Sirilli, Giorgio and Evangelista, Rinaldo (1998), 'Technological innovation in services and manufacturing: Results from Italian Surveys', *Research Policy* 27, 881–899.

Sundbo, Jon (1998), 'Standardisation vs. customisation in service innovations', *SI4S Topical paper* 3, Oslo: STEP Group.

Sundbo, Jon and Gallouj, Faïz (2000), 'Innovation as a Loosely Coupled System in Services', in Metcalfe, John S. and Miles, Ian (eds), *Innovation Systems in the Services Economy: Measurement and Case Study Analysis*, Boston: Kluwer Academic Publishers, pp. 43–68.

Tomlinson, Mark (2000), 'The contribution of knowledge-intensive services to manufacturing industry', in: Andersen, Birgitte, Howells, Jeremy, Hull, Robert, Miles, Ian and Roberts, Joanne (eds), *Knowledge and Innovation in the New Service Economy*, Cheltenham: Edward Elgar, pp. 36–48.

Windrum, Paul and Tomlinson, Mark (1999), 'Knowledge-intensive services and international competitiveness: a four country comparison', *Technology Analysis and Strategic Management* 11, 391–408.

Chapter 4

Innovation and Technological Change in Tourism: A Global-Local Nexus

Christian Longhi

Tourism[1] is inseparable from its spatial dimension. But the inscription of tourism in space has deeply evolved over time. The future represents a major challenge for the actors – public and private – involved in the activity. This chapter establishes that the geography of contemporary tourism is the result of complex and interdependent global and local dynamics. The constant reshaping of the local is simultaneous with the globalization processes of this and related industries (Salazar 2005). Obviously competition among alternative destinations is a relevant element, as is the secular decrease of the transport costs.[2] But the issue is not restricted to mere variations of elements in given markets. In the industry of tourism, technological changes and innovation have resulted in a new organization of the industry and markets, changes in interactions and interdependence between the different actors, the emergence of new institutional frameworks and rules, and ultimately changed territorial dynamics.

Globalization and developments in information and communication technologies (ICT) feed the present process of changes and shape the new landscape of tourism. The link between these two processes – innovation – works as the engine of the industry. Tourism is at the heart of these processes; it has been able to plug into and implement renewed knowledge bases, with innovative outcomes.

According Poon (1993) travel and tourism must be treated as an information-intensive industry. The internet is particularly useful for the industry. On the demand side, its forums reduce the uncertainty related to products, exerting an instantaneous control on the quality of information or products supplied. On the supply side, it allows rapid modification and diffusion of changes, in reaction to unforeseen events. Nevertheless, the uses of the internet do not only amount to the renewal of

---

1   The definition of tourists and tourism has raised much controversy. A tourist is defined as a person out of its usual main residence for at least 24 hours (or one night) and one year at most (World Tourism Organization). Tourism gathers all the activities aiming to satisfy the needs of the tourists; its main characteristic is to gather very heterogeneous elments, micro firms and multinationals, traditional and high tech activities.

2   The general findings of the New Economic Geography (Krugman, 1991) can be applied to tourism; the long term decline of transport costs can trigger a transfer from European destinations towards a peripheral destinations and contribute to the explanation of the current Europe stagnation for instance.

the relations between supply and demand, they allow above all the emergence of new relations *between* consumers and *between* suppliers (Gensollen, 2001). B2B and C2C relations are indeed the bulk of the changes. The process of formation and expression of demand have been radically transformed, as have the interactions and interdependences governing the industry.

This chapter aims to apprehend their impacts on tourism, with a special focus on the territorial dynamics in the European and French cases. It defines a relevant analytical framework to apprehend the basic effect of internet on the industry of tourism. A focus on systems of production and innovation will be shown to provide a relevant analytical framework to apprehend the different dimensions of the changes at work. The chapter traces the evolution resulting from the emergence and development of e-tourism, its impacts on the coordination of the activities and the markets, in the face of globalization. After considering the implications of the internet for mobility, the paper concludes that global and local dimensions are deeply interrelated in the tourism industry. The industry exhibits the main features of intensive, knowledge-based innovation.

Paradoxes

Tourism is a worldwide industry. The World Travel and Tourism Council claims that tourism has become the largest world's industry (WTTC 2006), with roughly 8.3 per cent of the world's total employment and 10.6 per cent of total GDP (US$ 623 billions of economic activity in 2004). This economic strength of tourism, combined with a strong potential for growth, has induced deep competitive processes and significant industrial reorganization.

Europe is by far the first tourist continent, with roughly 54.5 per cent of international tourist arrivals (417 million of 763 million in 2004). The direct and indirect impacts of tourism are expected to account for 11.5 per cent of GDP and 12.1 per cent of total employment in 2005. Nationally, France is the first world tourist destination, welcoming 75 million international tourist arrivals. In terms of receipts, the US leads (€ 64.3 billions), far ahead Spain (€ 39.6) and France (€ 36.6). Despite this evident strength, the sector's structural characteristics underline an important fragility. In France, roughly half of the employment is provided by firms with fewer than ten employees (97 per cent of hospitality firms have fewer than 10 employees), and 25 per cent of employment is unsalaried. Paradoxically, the leading rank of France goes with a seemingly archaic industrial context poorly equipped to face the contemporary challenges of global competition. However, other branches of the tourism industry are highly concentrated and act as global players, leading many knowledge-based innovations.

Despite the short terms fluctuations inherent to tourism, impressive rates of growth are expected in the medium term. In the year 2020, the WTO forecasts 1.56 billion international arrivals – 717 million in Europe – and corresponding receipts of US$2000 billion per year. Alongside tourism, ICT (and especially the internet) is the other fast growing industry of the western economies. Obviously, the two trends do not occur by chance and are closely linked and interdependent, in terms of market and business as well as in the innovation. ICT has indeed considerably impacted tourism (Buhalis 1998; Wade and Raffour 2000).

Tourism takes a large share of overall e-commerce activity. Many e-commerce technologies and business applications have been developed by tourism related actors. E-tourism amounts to 30 per cent of the activity in tourism in the US in 2004 and 7 per cent in France, where up to 50 per cent of the transactions made in e-commerce have been related to tourism.[3] Moreover, the share of e-tourism is growing massively. For instance in the US, income related to travel activities decreased by 8 per cent in 2001, and 4 per cent in 2002, while the incomes of online transactions increased by 45 per cent (to US$27 billion). In Europe and in France, the growth of e-tourism reached 50 per cent in 2003 and 2004, with 40 per cent growth expected in 2005, while tourism overall was facing declines. The most important French site, voyages-sncf.com, has grown by 75 per cent in 2003 and 71 per cent in 2004.[4]

Despite these empirical facts, innovation in tourism, as in services in general (Tether 2000), has received little attention. Following Pavitt (1984), all service activities have been gathered in the 'supplier dominated sectors'. After many works acknowledging the role of innovation in service, Soete and Miozzo (1989) have revisited the taxonomy to cope with their diversity. They have distinguished new categories of service businesses:

- Production-intensive, scale-intensive (transport and travel, distribution) and network services (telecommunication), in which technological innovation encourages the standardization of service outputs, or, in more sophisticated systems, the adaptation (through customization) of standard services to particular user needs;
- Specialized technology suppliers and science-based sectors, including software and specialized business services, laboratory and design services, in which the main source of knowledge and technology is the innovative activity of the services themselves (Tether and Metcalfe 2003).

The tourism industry is not monolithic. Tourism gathers all the activities dedicated to the satisfaction of the needs of the tourists. Its products are complex and heterogeneous, combinations of elements separated in time and space (Caccomo and Solonandrasana 2001), often pre-defined packages assembling interrelated products and services (transport, accommodation services, leisure services, etc.). This notion

---

3   e-tourism is associated to the commercial transactions made on the net to the related markets and organizations. It does not cover the whole usages of the Internet in tourism, which have a direct impact on the organization of the industry independently of their commercial nature (CtoC mediated communities and forums for instance).

4   www.journaldunet.com 08.07.2004, Les Echos 21.01.2005; the growth of the French sites averages to 50 per cent. The turnover of the online travel amounts to € 19.2 billions in 2004, and will rise to € 28.6 billions in 2005.

of packaging or bundling is the core of the activity. Tourist products and services are often experience goods: the quality or utility is not known *ex ante* by the consumers. A system of advices and critics is thus necessary to the formalization of choices (Gensollen 2003). Contrary to the traditional good sectors where resources are transformed to be delivered to the customers, in tourism the customers have to go to the resources. Whatever their intrinsic qualities, the resources acquire an economic value only with the organization of the traveling of the tourists and development of the activity (Spizzichino 1991). Tourism is deeply rooted spatially, and involves geographical and industrial elements.

According the seminal definition of Leiper (1979) tourism is an open system; its interactions and interdependences impose this analytical framework. 'The elements of the system are tourists, generating regions, transit routes, destination regions and a tourist industry. The five elements are arranged in spatial and functional connections. Having the characteristics of an open system, the organization of the five elements operates within broader environments: physical, cultural, social, economic, political, technological with which it interacts' (Leiper 1979, 404). There are additional interrelationships between public and private actors, the Destination Management Organizations (DMO)[5] (national, regional, local) and the industry. This chapter focuses on the *tourism industry* and eventually *destination regions*, as they have an important role in closing the loops in the whole system and subsume regional development.

The tourism industry can be considered as a sector because it involves directly or indirectly numerous actors or group of actors interrelated by a complex set of market and non market relations A sectoral system of production and innovation (SSPI) will be defined after Malerba (2001) as

> a set of new and established products for specific uses and the set of agents carrying out market and non-market interactions for the creation, production and sale of those products. The agents are individuals and organizations at various levels of aggregation with specific learning processes, competences, organizational structure, beliefs, objectives and behaviors. They interact through processes of communication, exchange, cooperation, competition and command, and their interactions are shaped by institutions (rules and regulations). Over time, a sectoral system undergoes processes of change and transformation through the coevolution of its various elements.

The SSPI can be characterized by a set of attributes: knowledge bases and learning processes, organizations (firms and non-firms, networks), institutions, demand and geographical boundaries. The application of the concept of SSPI to services has been done in Andersen et al. (2001) or Tether and Metcalfe (2003).

These authors emphasize the specifity of services, which in contrast with manufacturing tend to be defined in terms of their processes (e.g. retailing, transportation, financial intermediation) rather than their products. Services are

---

5   The Destination Management Organizations (DMO) are the organizations responsible for the management and/or marketing of destinations at the national, regional or local levels.

essentially relational; they are involved in multiple transformations that transcend any simple definition of a sector (Tether and Metcalfe, 2003). This specificity of the sectoral dimension of services is particularly keen in tourism, which cannot be reduced to a simple product. The definition of the tourism industry has raised many problems and controversies (Wilson 1998; Longhi 2003). According to Leiper (1979) the tourism industry comprises the firms which purposely undertake the *joint coordination of their activities* for the purpose of servicing tourists. The coordination of the activities is the basic issue, more than the product. The definition of the sector cannot result from the usual criterion of technological substitutability, but from the organizational complementarity and interdependence between actors and group of actors sharing the responsibility and planning tourists flows (Tremblay 1998). Consequently, the industry vertically links firms which associate their specific competences to design a coherent product through the synchronization of their activities in time and space (Tremblay 1999; Werthner and Klein 1999). Basically, the traditional value chain of the travel and tourism industry can be decomposed in five main types of actors :

- Primary suppliers and service providers; they are the transport companies – airlines, railways, shipping lines, rental cars – and the accommodation service providers.
- Global distribution systems (GDS), which have been originally created by the airlines companies. They have been (with banks) the first application of ICT worldwide. The American Sabre, Galileo and Worldspan and the European Amadeus are competing for world leadership. The GDS have made available worldwide their proprietary internal reservation systems to the travel agencies – their privileged customers – to offer direct access to the airlines flights and fares, but progressively to hotel reservation, rental cars, ferries, and the multiple services linked to tourism. They are virtual travel market places linking sellers (airlines, hotel, etc) and buyers (travel agents, for the final customers), where the whole coordination of the activity is solved.
- Tour operators, whose specific role in the traditional division of labor is to bundle the tourism products. Often, the tour operators partly book amounts of elementary products necessary for bundles to the suppliers, at preferential terms.
- Travel agencies, which distribute the different products (from services providers and tour operators) to the consumer. In the traditional system, they are the quasi-exclusive contact with the tourists ('order takers'), and they claim a role of main adviser directing the choices of the consumers. A very important share of their turnover results from the sales of tickets.
- Consumers (leisure or business tourism) at the end of the value chain.

This traditional architecture of tourism – the couple tour operator-agency which assembles and distributes packages in close cooperation with suppliers and GDS, with a well-established system of fixed commission rates and strict regulations – is definitely bygone. New actors, new strategies have instituted re-arrangements of the markets and the organization of the industry. It will be impossible to provide in depth analysis of each of the different elements of the value chain we have described. The focus of the analysis will be placed on the dynamics which disrupt the architecture of the industry and recombine the interactions of the elements, because they command the understanding of the nature of the globalization and the territorial dynamics it entails.

Low Cost and Regions

What are the effects of the internet on the tourism industry? The naïve belief that easily accessible infomediaries and shopbots comparing prices would lead to pure competitive markets thanks to perfect information and decreased information asymmetries has not lasted long. 'The Great Equalizer?' has indeed resulted in the usual mechanism of price discrimination (Brynjolfsson and Smith 2001; Clemons et al. 1998). Price customization is even more and more efficient: customers can pay higher of lower prices buying the very same item from the very same site at the same time (Feldman, Turow and Meltzer 2005; Ramasastry 2005). This process is at odds with the system of fixed commissions charged by the intermediaries on the service provider prices. The issue is thus not limited to the usual pure market mechanisms. The changes in the whole organization of the industry and of the markets have to be tackled to understand the emergence of such customization. They do not regard the improvement or the extension of established things than the emergence of the new. So in airlines and transportation services, it is not the emergence of perfect markets, but the emergence of low costs which is the basic consequence of the internet. The same could be shown for the online agencies. The entrants, sometimes qualified as 'barbarians' in the literature (Wade 2000), have in fact exploited unused growth potentials and recombined the markets, or developed basic organizational innovations.

Southwest Airlines in the US has been the first of these new 'low cost' companies, basing its strategy on the idea that consumers of travel are more sensitive to prices than to superfluous additional services. Point to point and online booking have been the solutions found by the low costs to reduce the distribution costs of the traditional airlines. The two elements are perfectly complementary: point to point allows the implementation of simple, efficient, and rapid sites where the consumers immediately find the requested service, and the very low prices[6] compared to the standard ones inclines to the adoption of this new service.

This model has been implemented with success in Europe after 1995 by easyJet, Ryan Air, Go Fly, Buzz, and Virgin Express. Rapidly, these new entrants have radically changed the intra-European air travel market. Different strategies are at work: easyJet tries for instance to have access to the main airports when Ryan

---

6   All variable costs have been minimized services, but also homogenous air flight (usually Boeing 737, but also Airbus320 or 319), often remote airports, lower wages.

Air chooses the provincial ones. These different strategies regarding location complement different commercial strategies. Ryan Air requests subsidies from the local governments of its destinations, in exchange of a commitment in terms of number of passengers (tourists) 'supplied' to the local economy, while easyJet does not use these policies. Traditional companies (Lufthansa, KLM) have attempted to create low cost subsidiaries to arrest market share loses; but have largely ended these experiments.

Today, the vast majority of low cost airlines' tickets are sold online. In Europe, and particularly in France, these companies have played a major role in promoting internet transactions. The internet enables direct access to consumers: it underlies the processes of disintermediation and deconstruction of the traditional value chains in tourism. The low cost airlines have thus inaugurated processes of reorganization of the tourism industry. Markets are no longer cleared the way they used to be. Existing interactions and interdependences have been overthrown.

This creation of new markets results in new price formation and evolution. In addition to the recombination of the sectoral system, this has had a profound impact on local development. The low cost airlines have recomposed the landscape of the intra-European airlines transport. The revival of the deserted regional airports they have induced has reshaped the tourism economy of many regions. Further, many of their customers are new air travelers, not the clients of the traditional airlines. Finally, the low cost airlines are changing the impacts of tourism on regional development. In the case of France specifically, a huge share of the low cost traffic (60 per cent in 2002) comes from UK (massively from the London area). It is specially addressed to the airports of Nice and Paris, but the phenomenon is widespread and concerns thirty airports.

England is a leading generating region of international tourists, using packages from the Tour Operators. The low cost airlines have diverted the consumers from the mass, standardized and rigidly packaged tourist products dedicated to given destinations, toward behaviors oriented towards individual arrangement of the products and new trends. This has boosted the acquisition of second homes in France by English and North European people, not only in the French Riviera, but also in Bretagne, Dordogne, and Limousin, revitalizing rural areas. These new flows have a strong impact on the local economies, reinforced by the benefits from the short stays (a passenger spends €500 on average; the impact of the English visitors in Limousin has been estimated at € 165 million in 2003 (Perri 2005)). The low costs' traffic is thus also a regional issue in France, motivating development strategies to promote destinations to the individual customer. The companies do not ask for favorable conditions for their operations in Nice or Paris, but their location in secondary airports is usually the result of some agreement with the regional or local government (such as the financing of national and international marketing of the region focused on the low cost potential market to boost the traffic, sometimes in contradiction with the competition policy). Some regions have implemented development strategies around these new opportunities, developing short term leisure practices for the visitors as well as innovative structural actions for the new residents induced by the low cost airlines availability. For instance, English versions of the local e-administration websites are not to be found in the large cities, but in rural areas to make easier the integration of the new residents. The low cost airlines have thus not only totally transformed European short haul travel. They have implemented a new business model which bypasses the intermediaries of the traditional system. This business model has diffused and resulted in the emergence of a new organization of the industry. They have also impacted the whole tourism system as defined in Leiper (1979), the generating and destination regions, and regional development fostered by tourism.

Internationalization and e-Globalization

The internet and e-tourism, even while representing relatively small market shares, have motivated major reorganization of Europe's tourism markets. They have totally changed the process of internationalization of the European industry. With the development of e-tourism, changes are not limited to the entry of new actors, but have transformed the role of all the actors of the system.

In the nineties, European tourism producers engaged in vertical and horizontal integration to cope with the processes of internationalization, increased competition, and mass production which characterized the western economies during this period. Size, breadth, and control over all aspects of tourism were seen as key to secure access to potential demand. Preussag (now TUI), Thomas Cook and Rewe have taken over most of their competitors, following the same industrial and financial strategy. Preussag, a steel company which diversified through the takeover of TUI, the German leader, has become a world leader in less than five years with the acquisitions of Thomson (UK), Nouvelles Frontières (France), and also the Spanish, Belgian, Danish, Finish leaders. This strategy of horizontal integration of the European tour operators has been coupled with a strategy of vertical integration. TUI, with roughly 200 million customers, includes about a hundred brands, about 4000 travel agencies, 300 hotels, 100 planes, four airlines (including Corsair and the low cost Hapag-Lloyd Express). TUI has acquired Havas Voyage, certainly the lead brand in tourism in France.

In 2000, however, the locus of the large firms' acquisitions turned to e-tourism.[7] This shift has been motivated by Europeans' increased use of the internet for

---

7 Obviously, TUI has adapted its strategy; the online activity of the group is steadily improved. TUI has recently created ultravances.com, a discount online agency which rests on the unsold of the subsidiaries of the group, from airlines to hotels. Still, it has played a secondary role in the general reconfiguration of the tourism system, when it could have played the same role than Cendant in the US for instance. Indeed Cendant, one of the US leader in tourism industry – the world's largest franchisor in hotel (Days Inn and Ramada Inn) and real estate (Century 21) businesses, which holds in addition car rental operators Avis and Budget, Cendant Travel, a network of traditional travel agencies, the interface hotel GDS Wizcom, the timeshare leader RCI – has bought the GDS Galileo, one of the leading GDS, and Trip.com,

information and transactions, and by the quickening pace of globalization in European tourist industries. While the share of e-tourism in Europe is far below the US, the forecast rates of growth are quite high. Globalization has induced important and complex processes of concentration. US leaders Cendant, Expedia and Travelocity have invested the European market, because of its salience in the market for tourism. Competition and cooperation evolve with the usages of the internet, resulting in the re-organization of the markets and the industry. These processes underlie the contemporaneous radical reconfiguration of the sectoral system of production and innovation (SSPI).

New interactions and interdependences have emerged. The development of e-tourism has broken the chain linking service providers and suppliers, GDS, tour operators, and travel agencies. The websites of the airlines and hotels are competing directly with the tour operators and the travel agencies (traditional or online). All the actors are on the same field.

In the pre-internet age, travel agencies were the only agents to have a ready access to flights and fares though their online proprietary GDS terminal. Customers' only alternative was to call each airline to complete a search. Internet sites have overthrown this system; flights and fares can be directly and easily seen and transactions performed, without travel agencies for the consumers and without GDS for the agencies. The quality of the information can be controlled using different sources, including C2C relations and virtual communities of experience where important amounts of information regarding the quality of the tourist products are shared by the consumers (Kozinets 1999; Kim et al. 2004; Wang et al. 2002). The customers are indeed transformed into 'consum-actors' (Raffour 2002), deeply involved in the production of their own customized packages. They stand as a full-fledged actor of the industry: they buy more and more directly from primary suppliers, building themselves their product packages. This behavior challenges the viability of the intermediaries (traditional or online), as it bypasses the whole business model of the tourism industry. The internet and the changes induced by the convergence of the technologies have allowed a process of re-intermediation, a new arrangement of the system in which the consumers are able to handle and exchange freely information which required formerly closed proprietary technologies (Treboul and Viceriat 2003). The easy exchanges of large quantities of data at low cost allowed by internet have deeply transformed the inter-firm relations underlying the existing value chains. The process of convergence has resulted in the creation of the dynamic packaging, an important innovation which stands as the key for the future of intermediaries. It is defined as the combination of different travel components, bundled and priced in real time, in response to the requests of the consumer or

---

its online travel agency, and invested in all the information systems associated to tourism. This process towards more online markets involvement is continually deepening, and though its involvement in the European market, Cendant has led the recombination of the whole tourism industry. TUI will be shown to be one of the European leaders after the four main groups build around GDS.

booking agent. It allows consumers to buy different services at once but still have the 'pay-one-price' convenience.

The consumers had before to visit different independent sites to build customized tourism products, register their personal information several times and make several payments, or they had to accept to buy pre-packaged products. With dynamic packaging technology, the lack of functionality has disappeared, and customers can build customized packages that combine their preferences with flights, car rentals, hotels, and leisure activities in a single site and a single price. Some sites have implemented 'package savings calculator' that shows consumers how much they are saving over buying the different elements of their packages separately.

The investments induced by the implementation of these innovative services are considerable, but have to be undertaken to secure the viability of the activity in the contemporaneous online market. Technological change and the related process of convergence have redesigned the dynamics of cooperation and competition underlying the market and the organization of industry in tourism. The GDS have driven, directly or indirectly, the recomposition of the industry. They have moved from the status of suppliers and technology providers of the travel agencies to one of their direct competitors, selling directly or indirectly (through their own agencies) to the final customers. The process of concentration going with this strategy has resulted in the emergence of GDS-agencies expressing the recombination of the whole sectoral system.

Four main corporate actors have emerged (Longhi 2005): one European and three American companies.[8]

- Amadeus has grown as a leading global distribution system (GDS) and technology provider, primarily through its reservation and solutions for passenger management systems. It has rapidly developed leadership in internet technologies dedicated to the travel and tourism industry through its subsidiary e-Travel. Amadeus's cooperation with complementary actors has allowed it to shift its focus from technology services to online distribution services provider. This is how it has faced the challenge of the internet and taken a leadership in dynamic packaging. It has acquired Opodo, which amounts to 18 per cent of the online European market; Optims, now the leading European supplier of IT services to the hospitality industry; Promovacances, a leading online travel company in France with core expertise in pre-packaged holidays; and other European online actors. This strategy, moving from technology provider to virtual agency is not specific to Amadeus. On the contrary, the GDS has followed the model of its US counterparts, below.

---

8   A fifth actor, the German TUI, the tour operator European leader, complements this group of dominant players in Europe regarding market shares; but regarding innovation and re-definition of products and markets, the four GDS-agencies are clearly leading the whole process.

- The GDS Sabre invented this model of 'GDS-agency' with the creation of the pioneer online agency Travelocity at the very emergence of e-commerce, complemented by the acquisition of GetThere.com, a specialist in the online travel business. It has created a new site in Europe, Odysia, and acquired the discount travel site lastminute.com to compensate for its lack of brand awareness in Europe.
- Cendant, a US leader in tourism industry, owns the GDS Galileo, CheapTickets. com (the leading seller of discounted leisure travel products online), Orbitz (the third US online travel agency) and HotelClub.com (a full service provider of online hotel reservation services). It has considerably expanded its presence in Europe though the acquisition of a leading European online travel group, ebookers.
- Expedia, the world's leading online travel agency, owns Hotels.com, a leading provider of lodging worldwide, and relies on the independent GDS Worldspan for flight and fare information. Expedia entered the French market through the creation of a joint venture with the SNCF, voyages-sncf.com, which has rapidly grown as the leader of the French e-tourism market. Expedia created its own site in 2004, expedia.fr, and has acquired the European Anyway.com (a specialist in online leisure travel services) and Egencia, (a specialist in corporate travel management services). Dynamic packaging is the strategy implemented by Expedia to dominate the European market.

The competition between the GDS-agencies is now dependent on the cooperation with specific large end-providers. For example, in the past Amadeus and Sabre offered basically identical airline services to travel agencies. Today Opodo (Amadeus) and Travelocity (Sabre) may have highly different contents to display to their customers. The differentiation can be quite substantial regarding hotel offerings. Inter-Continental Hotels has for instance recently withdrawn its relation with Expedia, privileging its own online reservations and a unique relation with Travelocity. Thus a major problem for the four large GDS groups will be to feed and diversify their offering of hotels.

The new forms of tourism which are developing – and will develop even more with the implementation of mobile technologies – increasingly seek tailor-made experiences. The DMO, the regional or local travel organizations, the professional associations of local/regional small firms could each have an opportunity to post their offerings worldwide though agreements with the new online agents and benefit from the severe competition to build large destination portfolios. Online access could become a vector of local development through the implementation of dedicated regional policies supporting the required investments. Dynamic packaging could reorient the flows of tourists though direct access to the customer, and divert the tourists from mass-customization in proprietary hotel chains strictly ruled by the global travel organizations in dedicated destinations towards an effective customization. Clearly the competition facing the European destination is more and more severe. The design of innovative tourism policies, carrying out new combinations, new institutional and organization arrangements to paraphrase Schumpeter, should foster development benefiting from the usages of internet and from the recomposed tourism industry. The design of these development policies implies public/ private strategies able to accommodate with activities dominated by small and very small firms, i.e. to build clusters 'of interconnected companies and institutions in' tourism (Porter 1998) or local tourism systems (Decelle 2004). The basic role of local competences in the knowledge-based activities in the globalization, largely documented in the high-tech industries (Longhi 2005), is even more important in the contemporary tourism, where the 'significance of market knowledge and procedural knowledge is highlighted' (Tether 2004).

The basic issue regarding tourism in Europe is the *decoupling* of the different elements of the tourism system defined in Leiper (1979). The implementation of clusters, of local innovative systems focused on the usages of the internet and taking advantage of the opportunities offered by new entrants, from low costs to online activities, is a necessary condition to work out the decoupling issue. Paradoxically, the emergence of mobile technologies is to emphasize the importance of coherent strategies of local development regarding the viability of tourism.

Restless Tourism

The emergence of mobile technologies is symptomatic of the 'restless' nature of tourism in the sense of Metcalfe (2002). The potentialities of mobile technologies in tourism are huge. The new technologies make it possible for travelers to be geographically independent and for social life to be networked. Besides the domains classically dominated by fixed-access internet, mobile technologies allow the packaging and marketing of tourist activities and needs decided *on site*. Mobile technologies are key to providing continuity of services through real-time access to information or services. Implementation of the new generation of mobile and wireless technologies – the so called 3G – will allow tourists to use services dedicated to localization, positioning, reservation, payment, and check in, via mobile phone.

Competition for client access and overall market organization are the key issues affected by mobile technologies. The boundaries of the system are not fixed, and can change over time as the knowledge base evolves and the related actors enter the system. Mobile phone operators will likely increase their investment in e-tourism. In any case, mobile technologies are the future of e-tourism and will impact the whole tourism system, as depicted in Leiper (1979).

These trends have already been embodied in Japan, where NTT DoCoMo – followed by its competitors – has developed solutions combining e-money and mobile phone (FaliCa technology) allowing customers to make online reservations, pay online, and check in at the airport with their phone. The system has been adopted by the train companies, the metro, and more and more businesses. Besides transactions, the technology is also used for passkeys into homes and businesses. These technologies are available in Europe, but the business model has to be defined before they can be used and implemented. The mobile phone will have to work as a device for secure payment, requiring complex organizational arrangements. The interrelationships between business models, organizational forms, technologies and outputs are brought to the fore in tourism, where markets do not pre exist, but are instituted in the competitive process and co-evolve in the competitive process (Metcalfe 2002).

The multiplication of services available through mobile communications will have a major impact on the tourism system, raising again the issue of decoupling of its different elements. Mobility will deepen the role of local development in tourism and the definition of consistent regional and local strategies. Services and infrastructures have to be designed to allow mobility, to allow a rambler or cross-country skier to enjoy nature with free access, if needed, to the services attached to the mobile technologies.[9] Public and private strategies will have to build service-oriented clusters to foster regional or local development from mobile technologies in tourism destination areas. Recomposition of the different elements of the tourism system (Leiper 1979) is certainly necessary in Europe to support the efficient implementation of such strategies.

Conclusion

The internet has revolutionized the sectoral system of production and innovation of the tourism industry. Indeed, the tourism industry incorporates the main features of the contemporary knowledge-based economies: globalization, permanent competition and innovation, mobility, and the usages of the internet are at the heart of these different processes. Evolving knowledge bases meeting the technological and information driven innovations related to the usages of the internet have resulted in new entrants, in changes in the interactions and interdependence, and in the boundaries between cooperation and competition prevailing in the tourism industry. The organization of the industry and its markets has been turned upside down, new regulations and modes of price formation have emerged, and more will emerge with the development of mobile technologies.

Europe is a field of competition fuelled by globalization in the contemporary tourism industry. The reorganization of the industry reorients generating and destination flows; the huge and deep changes induce the decoupling of the elements of the tourism system. The formation of clusters, local tourism systems relying on

---

9   More and more innovative clusters related to tourism are indeed emerging in Europe. The small ski resorts of the south French Côte d'Azur, where broadband internet has been deployed thanks to national and regional policies, are an example. Innovative technologies have been developed. The first low cost ski-lift passes has been developed in the small Valberg resort, www.Valberg-lowcost.com, introducing yield management in the lift passes, lower prices going with orders placed up to a month before the arrival in the resort. The implementation of the mobile technologies will have to go with new investments in infrastructures and new innovative marketing strategies.

cooperation among public and private actors to secure the investments necessary to the adaptation to the innovative uses of technologies, has been shown to be a key to foster regional development from the changes. The local/global nexus distinguishing the knowledge-based services of the contemporary economies is a basic feature of the tourism industry.

References

Andersen, B., Metcalfe, J.S., Tether, B.S. (2001), Distributed innovation systems and instituted economic processes, ESSY working paper Services Sectors in Europe, TSER DG12.

Brousseau, E. and Curien, N. (2001), 'Economie d'internet, économie du numérique', in Brousseau E. and Curien N. (eds), *Economie de l'Internet, Revue Economique*, n° hors série vol. 52 – Octobre.

Brynjolfsson, E.M. and Smith, D. (2001), 'The Great Equalizer? Consumer Choice Behavior at Internet Shopbots', Working Paper, December (Cambridge, MA: Sloan School of Management).

Buhalis, D. (1998), 'Strategic use of information technologies in the tourism industry', *Tourism Management* 19, 409–421.

Buhalis, D. (2004), 'EAirlines: Strategic and Tactical Use of ICTs in the Airline Industry', *Information & Management* 41, 805–825.

Buhalis, D. and Licata, M.C. (2002), 'The Future of eTourism Intermediaries', *Tourism Management* 23, 207–220.

Caccomo, J.L. and Solonandrasana, B. (2001), *L'innovation dans le Tourisme. Enjeux et Stratégies* (Paris, Coll. Tourismes et Sociétés, L'Harmattan).

Charbit, C. and Fernandez, V. (2003), 'Sous le Régime des Communautés: Interactions Cognitives et Collectifs en Ligne', *Revue d'Economie Politique*, n° hors série 113, 229–252.

Clemons, E.K., Hann, I.H. and Hitt, L.M. (1998), *The Nature of Competition in Electronic Markets: An Empirical Investigation of Online Travel Agent Offerings*. Working Paper, June. The Wharton School of the University of Pennsylvania.

Decelle, X. (2004), 'A Conceptual and Dynamic Approach to Innovation in Tourism', Proceedings of the Conference on Innovation and Growth in Tourism, Lugano, Switzerland, September 2003.

Direction du Tourisme (2005), *Les chiffres clefs du tourisme*. http://www.tourisme.gouv.fr.

E-tourisme newsletter (2006), http://www.etourismenewletter.com.

Feldman, L., Turow, J. and Meltzer, K. (2005), *Open to Exploitation, American Shoppers Online and Offline* (Philadephia: Annenberg Public Policy Center, University of Pennsylvania).

Gallouj, F. and Weinstein, O. (1997), 'Innovation in Services', *Research Policy* 26, 537–556.

Gensollen, M. (2001), 'Internet: Marché électronique ou Réseaux Commerciaux', *Revue Economique*, n° hors série 52, 191–211.

Gensollen, M. (2003), 'Biens Informationnels et Communautés Médiatées', *Revue d'Economie Politique*, n° hors série 113, 9–40.

Kim, W.G., Lee C. and Hiemstra, S.J. (2004), 'Effects of an Online Community on Customer Loyalty and Travel Product Purchases', *Tourism Management*, à paraître.

Kozinets, R.V. (1999), 'E-tribes and Marketing: Virtual Communities of Consumption and Their Strategic Marketing Implications', *European Management Journal* 17: 3, 252–264.

Krugman, P. (1991), *Geography and Trade* (Cambridge: MIT Press).

Lafferty, G. and van Fossen, A. (2001), 'Integrating the Tourism Industry: Problems and Strategies', *Tourism Management* 22: 1, 11–19.

Leiper, N. (1979), 'The Framework of Tourism: Towards a Definition of Tourism, Tourist, and the Tourist Industry', *Annals of Tourism Research* 6: 4, 390–407.

Leiper, N. (2000), 'An Emerging Discipline', *Annals of Tourism Research* 27: 3, 805–809.

Longhi, C. (2003), 'Quelle Offre? Des Acteurs – Opérateurs Composites', in J. Spindler (ed.), *Le Tourisme au XXI Siècle* (Paris: L'Harmattan).

Longhi, C. (2005), 'Local Systems and Networks in the Globalisation Process', in *Research and Technological Innovation, The Challenge for a New Europe*, A. Quadrio Curzio and M. Fortis (eds) (Heidelberg: Physica Verlag).

Longhi, C. (2005), *Usages of the Internet and e-Tourism. Towards a New Economy of Tourism*, (Gredeg: CNRS, Sophia Antipolis).

Malerba, F. (2001), 'Sectoral Systems of Innovation and Production: Concepts, Analytical Framework and Empirical Evidence', ECIS Conference 'The Future of Innovation Studies', Eindhoven, September 20–23.

Malerba, F. (2002), 'Sectoral Systems of Innovation and Production', *Research Policy* 31, 247–264.

McAfee, R.P. and Hendricks, K. (2003), 'Evolution of the Market for Air-Travel Information', mimeo, University of Texas at Austin.

Metcalfe, J.S. (2000), 'Co-Evolution of Systems of Innovation', Volkswagen Foundation Conference on Prospects and Challenges for Research on Innovation, Berlin, June 8th-9th.

Noyer, O. (2004), 'Les GDS, de la Régulation à la Dérégulation', *Les Echos* 19267, Dossiers Transports, 18 Octobre.

OECD (2004), Introduction to the Proceedings, Conference on Innovation and Growth in Tourism, Lugano, Switzerland, September 2003.

Origet du Cluzeau, C. and Viceriat, P. (2000), *Le Tourisme des Années 2010: La Mise en Futur de L'Offre* (Paris: La Documentation Française, http://www.tourisme.gouv.fr).

Penard, T. (2002), 'Mythes et Réalités du Commerce Électronique: Une Revue des Études Empiriques', in M. Basle and T. Penard (eds), *e-Europe: la société européenne de l'information en 2010* (Paris: Economica).

Perri, P. (2005), *Impact des Compagnies Aériennes Low Cost sur les Prix de L'Immobilier: Cas du Limousin, de la Dordogne et de l'Aude* (Paris: Direction du Tourisme).

Poon, A. (1993), *Tourism, Technology and Competitive Strategies* (Wallingford: Cab International).

Porter, M.E. (1989), *The Competitive Advantage of Nations* (New York: Free Press).

Porter, M.E. (2001), 'Strategy and the Internet', *Harvard Business Review*, March, 63–78.

Raffour, G. (2002), *L'impact des Nouvelles Technologies de l'Information et de la Communication dans le Secteur du Tourisme. Enjeux et Recommandations.* Rapport au Conseil National du Tourisme, http://www.tourisme.gouv.fr.

Ramasastry, A. (2005), 'Websites that Charge Different Customers Different Prices: Is their Price Customization Illegal?' *Finlaw*, June, 20.

Spizzichino, R. (1991), *Les Marchands de Bonheur: Perspectives et Stratégies de l'Industrie Française du Tourisme et des Loisirs* (Paris: Dunod).

Tether, B.S. (2004), *Do Services Innovate (Differently)?* CRIC Discussion Paper 66, University of Manchester, November.

Tether, B.S. and Metcalfe, J.S. (2003), *Innovation Systems and Services. Investigating 'Systems Of Innovation' in the Services Sectors – An Overview.* ESSY working paper and CRIC discussion paper n°58, CRIC, University of Manchester, February.

Treboul, J.P. and Viceriat, P. (2003), *Innovation Technologique dans les Produits et Services Touristiques*, Direction du Tourisme, Mars.

Tremblay, P. (1998), 'The Economic Organization of Tourism', *Annals of Tourism Research*, 25: 4, 837–859.

Wade, Ph. (2001), *Impact des Nouvelles Technologies sur les Systèmes D'information et de Réservation Touristique*, Rapport au Conseil National du Tourisme, http://www.tourisme.gouv.fr.

Wade, Ph. and Raffour, G. (2000), *Tourisme et Technologies de L'information et de la Communication: Le Futur Est Déjà Là*, Rapport au Conseil National du Tourisme, http://www.tourisme.gouv.fr.

Wang, Y., Yu, Q. and Fesenmaier, D.R. (2002), 'Defining the Virtual Tourist Community: Implications for Tourism Marketing', *Tourism Management* 23, 407–417.

Wilson, K. (1998), 'Market/Industry Confusion in Tourism Economic Analyses', *Annals of Tourism Research* 25, 803–817.

PART 2
Internationalization of Service Firms

Chapter 5

Spatial Divisions of Expertise and Transnational 'Service' Firms: Aerospace and Management Consultancy

John R. Bryson and Grete Rusten

There have long been service economies – economies in which service employment has made a substantial contribution to economic competitiveness. The history of service employment is one of concentration in key urban places. This is reflected in the emphasis placed by Walther Christaller on the 'range of a good or service' in the delimitation of central places and the ranking of cities ([1933] 1966). The range is the maximum distance over which people will travel to purchase a service or a good; at some distance from the point of supply people will decide that the inconvenience of travel outweighs the value of the service or they will use an alternative center. It would be a truism to argue that the world has altered since Christaller worked on Southern Germany; three important related alterations have occurred – developments in information telecommunications technologies (ICT); the onset of globalization and on-going relentless developments in the technical division of labor.

'Central Places' and 'Global Cities'

Christaller was concerned with central places: places that had concentrations of economic actors and that were simultaneously places of power, production and consumption. The focus of much academic research is still on central places, key global cities that are considered to be information/expertise/knowledge rich places and are the centers of command and control over the developing global economy (Sassen 2001; Taylor and Walker 2001). The language alters but the arguments that are constructed resonate with similar debates and processes. In 1915 Geddes coined the term 'world city' (Clark 1996), in the early 1990s Daniels (1991) noted that 'higher order cities' have comparative advantage as centers of service industry and technology and in 1991 Sassen developed the term 'global city'. The global cities have been identified as the primary locations for advanced professional services; those services that provide expertise as an input into a variety of different production processes.

It is intriguing to juxtapose Christaller's conceptualization of central places with that being developed in the global cities literature. Christaller's understanding of the geographies of central places was founded upon an explanation grounded in consumption behaviour related to travel patterns. In contrast, the global cities literature rests on a set of assumptions regarding production and consumption. On the one hand, the growing service intensity in the organization of all industries has led to an increase in demand for a complex array of producer service functions – finance, insurance, real estate, legal services, accountancy, advertising, management, public relations, industrial design and a range of technical inputs. The increase in demand includes both intermediate consumers (private and public sector) as well as final consumers. On the other hand, Sassen (2001; 2002) and others have argued that the increasing service intensity of production systems has had significant growth impacts for cities from the 1980s into the present (Sassen 2002, 21). The actual start date is not material to our argument, but it is worth noting that the shift towards 'service intensity' has a much earlier origin; the 1980s date partly reflects the well-documented externalization of service functions in which in-house employees were replaced by external service providers (O'Farrell et al. 1993, Rusten 2000; Bryson et al. 2004).

The global cities literature has predominantly concentrated on understanding producer services in London, New York and Tokyo (Warf 2000). The argument is that there is something distinctive about these places as they provide advanced or world class expertise compared to other places. There is no question that London, New York and to a lesser extent Tokyo play extremely distinctive roles in the developing world economy. What is distinctive is the embeddedness of global financial institutions in these cities and especially specialist institutions that play an important role in forming, making and regulating financial markets. In many instances, these financial institutions entail wealth creation via manipulating flows of financial capital that have little if anything to do with local economies and have become largely disengaged from the wider production process. A significant proportion of the producer service activities in the global cities have nothing to do with manufacturing or are only indirectly involved through mergers and acquisitions in which the primary focus is creating wealth via buying, selling and restructuring firms rather than on producing goods. The difficulty is that the literature has become fixated on these 'higher order' financial services and, in some cases, uses them as proxy measures to identify the global nature of a city's activities or even to define them. Growth in service employment and numbers of firms has occurred in a range of different sizes of cities and not just the global cities (Rusten et al. 2004). The point is that the growth in services is occurring in cities at different levels in the urban hierarchy, and that 'some of these cities cater to regional or sub-national markets; others cater to national markets and yet others to global markets' (Sassen 2002, 22). Outside key urban centers, for example small market towns, growth in producer service employment has been relatively restricted and linked largely to local demand. Many of these firms provide standardized services, but specialist expertise can develop to support concentrations of economic activity.

The difficulty with the revival of interest in world or global cities is that the research focus has tended to be on understanding the dynamics of the premier league of cities to the neglect of detailed research into the role played by lesser cities (Markusen et al. 1999; Scott 2002); second cities or even large complex city regions that are the location for significant numbers of firms of all types (manufacturing to professional services), but are not tied directly into the global financial system as they are not home to one of the major stock or commodity exchanges (Daniels and Bryson 2005; Bryson and Daniels 2006). The global cities literature has moved beyond the confines of the premier cities by exploring ways in which other places are linked into the global city network. Conceptualising a network of globalized urban centers (Savitch 1996; Sassen 2001) expands the definition of global cities to embrace other key 'global' urban arenas. It also begins to explore the relationship between centralization of activities within key urban centers compared to the dispersed nature of production (Sassen 2002). This shift in emphasis towards networks of interaction provides a valuable opportunity to adjust the research focus away from the global cities as the primary site for analysis towards a return to the emphasis placed by Massey on 'the interpretation of the spatial organization of the social relations of capitalist production' (1984, 6). The problem is that the global cities literature has a tendency to spatial fetishism or in this case 'city or urban fetishism'. By this we mean that the starting point of the analysis or the dependent variable is the city and in research designs that may over-emphasis city-based elements of economic activity at the expense of understanding the role played by other places in the developing global economy.

We do not dispute that the global cities are important places but it is difficult to understand their role in the global economy without exploring the wider organization and geographies of producer service production more generally. The starting point of any analysis should be on the organization, geographies and developing division of labor of producer service firms rather than on the global cities. In this argument it is important to differentiate between financial services and other forms of producer or business services. The focus of the argument excludes financial services as these are heavily grounded within the key global cities. Nevertheless, we argue that other forms of service expertise (accountancy, legal services, design etc) are produced and delivered in complex ways that sometimes include and sometimes exclude the premier global cities. Our argument also does not differentiate between service expertise that lies within large manufacturing companies and that which lies with independent supplies of business and professional services (Daniels and Bryson 2002).

The Division of Labor

In 1776 Adam Smith published *The Wealth of Nations*, in which he established the foundations of economic theory and formulated the concept of a division of labor. Smith placed considerable emphasis on the importance of this concept by stating in the opening sentence that:

[t]he greatest improvement in the productive powers of labor, and the greatest part of the skill, dexterity, and judgement with which it is anywhere directed, or applied, seem to have been the effects of the division of labor (Smith [1776] 1977, 109).

Smith argued that the effects of the division of labor are more easily understood by working through examples of particular forms of manufacturing. He noted that the development of a division of labor is most apparent in 'trifling' sectors of the economy that are dominated by small firms as it is possible to observe the organization of a workforce that has been collected together in the same building. In large firms it is impossible to collect the workforce into a single factory and '[t]hough in such manufactures, therefore, the work may really be divided into a much greater number of parts than in those of a more trifling nature, the division is not near so obvious, and has accordingly been much less observed' (Smith [1776] 1977, 109). This is an extremely important point as the inference is that analysts may be more aware of visible forms of the division of labor and less aware of other forms.

There are at least two explanations for the existence of a division of labor. First, Smith provides an economic rationale by reasoning that the increase in the quantity of work that results from the division of labor is the product of three factors. First, the dexterity of employees improves and this increases their speed. Second, time is lost when an employee changes from one type of work to another; tools must be found, the working area altered, the worker might have to move to another part of the factory and the worker has to adjust to a new task. Third, machines permitting cheaper and faster production could be introduced to replace workers. It is through a division of labor that distinct parts of the production process become visible, and as they become visible machines can be developed to replace labor.

Second, Emile Durkheim provides a complex sociological explanation that is based upon a shift from mechanical solidarity in which individuals in a society differ from one another as little as possible on the basis of a shared experience of the world. This type of solidarity is only possible given the low level of the division of labor; it is a solidarity based on likeness and to Durkheim, in its extreme form, negates individuality such that a peasant is a peasant is a peasant. The opposite form of solidarity, organic solidarity, is one in which consensus comes from differentiation. It is organic as the organs of a living being each performs a different function, but each is an essential element of the total organism. As labor is divided each person experiences the world differently; each person develops there own 'sphere of action that is peculiar to him (sic), that is, a personality' (1933, 131).

The division of labor develops as populations grow and resource pressures begin to develop. The first step in the creation of a division of labor is the establishment of monarchy. The division of labor is driven forward in response to the increasing intensity in the struggle of individuals to exist; specialization of production enables individuals to survive by capitalising on their skills and expertise and differentiating themselves in the society. Durkheim's differentiation of occupation should not be confused with that of Smith as for Durkheim the division of labor is an expression of social differentiation and of the overall structuring of a society of which an economic division of labor is only a part. Durkheim's division of labor produces solidarity and social order as:

> [I]n the same city, different occupations can co-exist, without being obliged mutually to destroy one another, for they pursue different objects. The soldier seeks military glory, the priest moral authority, the statesman power, the business man (sic) riches, the scholar scientific reknown. Each of them can attain his (sic) end without preventing the others from attaining theirs. It is the same even when the functions are less separated from one another . . . Since they perform different services, they can perform them parallelly (Durkheim 1933, 267).

In his analysis, Durkheim began to speculate about the development of a 'spatial' division of labor (1933, 268-269). This is important as these ideas predate the development of a 'spatial division of labor' in geography by 91 years. He does not use the words 'spatial' or even 'geography', but instead speculates over what would happen when a region with a specialist product begins to compete with other places as a consequence of improvements in communications. He argued that '[s]imilar occupations situated on different points of land are as competitive as they are alike, providing the difficulty of communication and transport does not restrict the circle of action'. One consequence is that companies that are out-competed must either disappear or transform with transformation leading to greater specialization; the out-competed firm shifts position within the market so that it supports rather than competes with the more competitive firm. In some cases, the small firm will be subsumed by the larger with the small employer becoming a foreperson and the small merchant an employee. In other instances, 'two similar enterprises establish equilibrium by sharing their common task' (Durkheim 1933, 270). The latter case, for example, could represent the establishment of close working relationships between firms located in different places that might eventually lead to the creation of a global firm that consists of linked independent partnerships, as is the case for the large accountancy practices.

A Spatial Division of Expertise

In 1984, Doreen Massey published a critique of traditional economic geography in which she argued that the spatial organization of production had been transformed to such an extent that 'the old theoretical structures have lacked the flexibility to respond' (1984, 3). On the one hand, traditional theory had focussed on understanding individual firms while, on the other hand, the increasing importance of large, multi-plant companies made it imperative that analysts began to explore the behaviour of firms of different types. Massey was arguing for a shift in focus onto the *social organization of production* (Massey 1984, 4) and particularly the spatial organization of production. Massey's theoretical framework is applied to industry; defined as all economic activity – primary, secondary and tertiary. Part of this argument accepts that the economy is in a constant state of transition. Manufacturing in particular places is no longer as important as it was, and perhaps, more importantly, is the on-going shift towards white-collar or various forms of service work. Since Massey published 'Spatial Divisions of Labour' in 1984 academic debate as well as the economy has shifted away from debates grounded in traditional constructions of labor towards literatures that have posited the existence of a 'service class' (Goldthorpe 1982; 1995), 'creative class' (Florida 2002) and for a new geography of talent (Florida 2002b). This is to position this argument within wider debates regarding the rise of informational capitalism (Castells 1996) or the development of a service world (Bryson et al. 2004) or service-led economy (Marshall and Wood 1995).

This shift raises a number of important conceptual issues. It should not be assumed that the social organization of service organizations is similar to that of manufacturing. In fact, there are important differences, for example, in the nature of service innovation (Bryson and Monnoyer 2004), in the factors of production in which capital is less important than an individual's reputation and expertise or a firm's brand and in the legal organization of firms. The majority of producer service firms are not companies *per se* but rather partnerships. A partnership is a relation between two or more people carrying out business in common with a view to sharing profits rather than turnover and expenses. A limited company is an independent 'entity' in its own right whilst a partnership is not; a partnership does not have a legal identity separate from the partners. This means that in a partnership all the partners are jointly liable for all partnership debts to the full extent of their assets and all partners are self employed. In a limited company directors are not liable as long they have been acting with due diligence. The importance of partnership forms of business amongst producer services means that these companies have the potential to act in very different ways to limited companies. All partners are able to commit the partnership to a legal liability to a third party and all partners are jointly exposed to third party liabilities. The implication is that within a limited company a hierarchical management structure can be developed whilst within a partnership structures must be in place to ensure that all the partners are involved in the decision-making process.

These differences imply that there are dangers associated with applying terms that have been developed to explain a particular set of economic relationships to different forms of work and organization. The concept of a division of labor is rooted within manufacturing; Adam Smith's famous example of pin-making is one indicator of its association with manual labor. Within geography, the concept is traditionally associated with deskilling (Braverman,1974; Hoddos 2002), branch plants of manufacturing companies and the on-going development of an international division of labor or global shift (Dicken 2003) which is closely associated with textiles and clothing, automotive, electronics and to a lesser extent services. It is timely to reconsider the language that is used to describe economic processes. The term labor is politically loaded as it is closely associated with trade unionism as well as being strongly associated with particular theoretical positions, for example Marx's labor theory of value. In his early writings Karl Marx associates the working through of the division of labor with deskilling and dependency, for example:

[A]s a consequence of this division of labor on the one hand and the accumulation of capitals on the other, the worker becomes more and more uniformly dependent on labor, and on a particular, very one-sided and machine-like type of labor. Just as he (sic) is depressed, therefore, both intellectually and physically to the level of a machine, and from being a man to a stomach, so he also becomes more and more dependent on every fluctuation in the market (Marx, [1844] 1981, 285).

It is useful to begin to disassociate particular type of high added value service work from the associations that are wrapped through the term labor. Florida has drawn attention to the importance of the three 'Ts' of technology, talent and tolerance. (2002 a & b) whilst Bryson (1997) has emphasized 'reputation, embodied expertise and knowledge'. Castells argues that the restructuring of firms and organizations and developments in information technology 'is ushering in a fundamental transformation of work: *the individualization of labor in the labor process* (Castells 1996, 265, italics in original). Ultimately, this means that the spatial division of labor literature needs to reflect the increasing centrality of embodied expertise in the economy (Wellington and Bryson 2001; Bryson and Wellington 2003). Fundamental to these approaches is an appreciation that there has been a shift in the nature of particular forms of producer service work. This shift transforms work away from labor that is controlled by capital and regulated by company law to a world of work in which partnership forms of governance are important and one in which control or more correctly power in the 'employment' relationship is transferred from the employer to the employed. In manufacturing, owners of capital are able to engage in a complex game of global monopoly in which most of the power in the relationship between worker and employer is held by employers. The situation is much more complex with the development of expertise intensive occupations as the nature of embodied expertise transfers power from employers to employees. Indicative of this transfer is the difficulties experienced by firms in managing business service professionals, the importance of partnership forms of firm organization in this sector of the economy and the role played by bonus payments in staff retention. The most important asset 'owned' or managed by a business service firm is its staff and their embodied expertise; this walking highly mobile resource leaves the firm each evening and there can be no certainty that it will return next morning.

To draw attention to the distinctive features of specialist service providers it is desirable to shift away from the term 'labor' towards 'expertise'. In doing this, it is recognized that expertise-intensive service workers engage in 'labor' but that the nature of their labor, employment conditions, employer-employed power relationships and legal governance structures are distinctly different to that of other sectors of the economy. Labor cannot be removed or even disassociated from the production of service expertise, but the nature of labor has altered to revolve around the production and delivery of a series of intangibles – expertise, knowledge, creativity – that are embodied in laborers (service workers). In many respects, the nature of work experienced by people working in this economic sector is similar to that experienced by academics – a form of self-employed employment. This is especially the case for partners and self-employed professionals, but it must be acknowledged that below this level traditional employment relations exist in this economic sector.

The ongoing division of labor within the economy can be conceptualized as the workings through of a technical division of labor, henceforth expertise, or the development of an extended division of expertise (Sayer and Walker 1992). The development of a spatial division of expertise has meant that the consumption of goods and increasingly services in many countries relies on actual physical production elsewhere – in China, India and in other low cost countries. This means that some places increasingly specialize in the production of expertise; activities that are sometimes considered only part of the supporting architecture of the production system.

Not all service expertise needs to be developed and consumed locally. This means that different types of expertise are produced in different places. A spatial division of expertise exists within and between international business service firms. This means that particular branches of a global firm will specialize in particular economic sectors or processes. Clients requiring this type of expertise access it through their local office, and may never become aware of the fact that the expertise lies outside their regional economy. Our argument is that a *spatial division of expertise* is evolving that is part of immensely complex dynamic systems of production and that central to these systems is producer service expertise that either lies within or external to client companies. The building blocks of this 'expertise' economic system are distributed in a complex mosaic that reflects the ways in which a range of service firms (micro to large transnational firms) try to maximize profitability. The spatial division of expertise is informed by the social division of labor that for producer services lays considerable emphasis on relationships between people and on the joint-supply and co-production of service knowledge/expertise.

In the next sections of this chapter we turn our attention to exploring two contrasting examples of the operation of a spatial division of expertise. The selection of examples is deliberate and is designed to demonstrate that the on-going development of a spatial division of expertise is occurring in both manufacturing as well as service industries. The construction of the concept of a spatial division of expertise moves the focus of analysis away from the global cities to other places in which expertise is produced and consumed.

Spatial Divisions of Expertise within Boeing and Airbus

It is becoming increasingly difficult to distinguish between manufacturing and service companies (Daniels and Bryson 2002). Many physical products contain embedded services, for example computers, or are supported by service agreements. Jet engine manufacturers sell their engines cheaply and try to obtain service contracts and supply spare parts; these activities now comprise around 60% of revenues. The business models of companies like Rolls-Royce and General Electric front-load costs and back-load revenue generation (Gapper 2006). This produces a very distinctive business model that is linked to one of the longest product cycles; aircraft last for over 30 years and it can take over ten years for an engine manufacturer to recoup the development costs of a new engine. This means that jet engine manufacturers have a complex intra-firm division of expertise formed around research and development, production and a distributed network of service centers. For investors, the jet engine product cycle is related to periods of declining revenues related to substantial R&D expenditure that is followed by a long-term revenue stream. In the early 1980s, Pratt & Whitney decided not to become involved with the development of a new engine for the Boeing 737 and since this time the company has avoided committing substantial sums on new engine development (Gapper 2006). This has left the market to Rolls-Royce and General Electric while Pratt & Whitney's revenue stream has been closely related to servicing and supplying spares for its existing engines. This is a short-term business strategy; towards the end of the thirty year product cycle the company's revenue stream is beginning to wither away. Pratt & Whitney is able to compete to service and supply spare parts for old engines that are no longer protected by patents. Manufacturers are much more rigorous in protecting their intellectual property in new engines and also in locking customers into long-term servicing and spare part agreements. This example of the commercial jet engine business highlights the complex interweaving that has occurred between three types of business model: exploitation of R&D; production based on orders and long-term service agreements. To remain successful over the long-term a manufacturer cannot avoid the development costs of a new engine (over $15 billion) or the management and development of a network of engine service centers that are located close to their key clients.

The increasing symbiosis that is developing between manufacturing and services is changing the geographies of production in complex ways, creating new business models that capitalize on service/manufacturing expertise and also repositioning expertise as a key source of competitive advantage. There are many examples, but perhaps one of the most dramatic is found in the aerospace industry and especially the corporate struggle that is occurring between Boeing, the former market-leader, and Airbus, the European consortium that was established in 1970. The word dramatic refers to both the intensity of the on-going struggle for market leadership and also to the size and complexity of their products. In 2002, Airbus had captured 44% of the market for commercial aircraft and by 2004 this had risen to 53%. This does not mean that Airbus will retain this position as both companies are developing new aeroplanes using very different business strategies.

The Airbus strategy is to develop a super-jumbo double-decker jet, the A380, an aeroplane that will carry 555 passengers up to 8,000 nautical miles at half the speed of sound. The A380 will fly passengers to key airport hubs where passengers will transfer to smaller aeroplanes for the final part of their journeys. Boeing's strategy is built around the 787 Dreamliner that will carry between 200 and 300 passengers with a range of 8,500 nautical miles. The Dreamliner will be able to fly between New York and Tokyo without refueling and fuel consumption will be 20% less than aeroplanes of a comparable size. These very different aeroplanes are being developed and manufactured using different economic geographies and distinctive spatial divisions of expertise.

Boeing's 787 Dreamliner Project

For the Dreamliner project, Boeing has created a new spatial division of expertise that is constructed around capitalizing on Boeing's internal expertise combined with that of its partners that are distributed around the globe. This expertise is a combination of research and development with industrial design inputs; the latter being a core, but much under-researched, producer service. This expertise has developed in centers of excellence that support high-tech engineering activities and in places that are far removed from the triad of the global cities. It would be possible to argue that these places are the 'global cities' for the aerospace industry, but this would take the analogy too far; these aerospace places should be considered as core centers of excellence that are part of the global architecture of a key international industry.

Just under 4000 engineers are directly employed by Boeing in the design of the Dreamliner with the majority of these positions based at the company's Everett plant (Washington, USA) (Gates 2005a). Another 670 engineers are employed by Spirit Aerosystems, Wichita (Kansas, USA); Spirit used to be part of Boeing but was sold in June 2004 to Onex, a Canadian investment firm. Boeing is responsible for the overall design of the plane, but detailed design work has been transferred to a network of global partners. This is an extremely important development as it enables Boeing to share development costs, increase the size of its design team by drawing upon the resources of its partner companies (Gates 2005b) and also to market the product as having being partially designed and manufactured in it key target markets (Table 5.1). The Dreamliner will be assembled at Boeing's Everett plant and key components will be shipped or flown to Washington in modified 747 cargo freighters.

The Dreamliner project represents a complex spatial division of expertise that is intended to reduce the cost of the new plane as well as provide Boeing with a marketing advantage in key target markets. According to Lane Ballard, a Boeing employee working on the Dreamliner's Wing Life Cycle Project Team 'you can only be a world-class producer in so many areas. Other companies are much more lean and nimble at producing individual components. Their overheads and infrastructure is less [than Boeing's]' (quoted in MacMillan 2005). Of the plane's structural parts only the vertical fin will be made in Washington while the rudder and leading edges come from China, tail cones from South Korea, flaps from Australia, midfuselage section and wings and wing box from Japan, midfuselage sections from Italy, landing gear from England and the nose section from the USA (Table 5.1). This is both a conventional spatial division of 'production' labor and a spatial division of 'embodied' expertise. At Everett, Boeing has established a *Global Collaboration Center (GCC)* that functions as the central node in a global network of design inputs into the Dreamliner project. The GCC enables lead engineers based in Washington to engage in co-designing the 787 with engineers based in Wichita, Australia, Japan,

Table 5.1   The spatial division of expertise for Boeing's 787 Dreamliner project

Country	Company	Products	Engineering jobs
United States	Spirit (Wichita, Kansas)	Nose Section (Wichita), engine pylons (Wichita), fixed leading edges (Tulsa, Okla.), moveable leading edges (Tulsa, Okla).	670
	Vought	Rear fuselage sections (Charleston, S.C.)	100 (Charleston) 300 (Dalas)
	Goodrich Aerostructures (Chula Vista, Calif.)	Nacelles	160
	Boeing (Frederickson, Pierce County)	Vertical fin	95
	Boeing (Everett, Washington)		3600
Canada	Boeing Canada (Winnipeg)	Wing to body fairing assembly, aft pylon fairings, wing to body fairings, main landing gear doors (body) main landing gear doors (wing)	60
England	Messier-Dowty	Main landing gear, nose landing gear	30
France	Latecoere	Aft passenger doors, forward passenger doors	NA
Sweden	Saab	Aft cargo door, forward cargo door	NA
Italy	Alenia	Midfuselage sections, horizontal stabilizer	NA
China	Chengdu Aircraft Industrial Group	Rudder	NA
	Shenyang Aircraft Group	Vertical Leading Edge	
	Hafei Aviation Industries	Wing to fairing panels	
South Korea	South Korean Air	Wing tips, tail cone	NA
Japan	Kawasaki Heavy Industries	Midfuselage section, fixed trailing edge	190
	Fuji Heavy Industries	Center 'wing box' fuselage section	130
	Mitsubishi Heavy Industries	Wing box	250
Australia	Boeing Hawker de Havilland unit	Moveable trailing edges, in board flaps	80

Note: Number of engineers are projections for the end of 2005 made by Boeing's first-tier suppliers and may not include all engineering specialities. Production workers are not included.

Source: After Gates 2005a.

Italy, Canada, Russia, China and other centers across the USA. The technology enables concurrent design changes to be made to the aircraft in real time and the collaborative technology provides the Dreamliner's spatially dispersed design team with a virtual collective workspace. As part of this strategy, Boeing established the Boeing Design Center (Moscow) to capitalize on Russian expertise (Gates 2005a). This center employs 140 engineers who provide design support to Wichita and Everett as well as inputting into the design modifications required to transform the 747s into the large cargo freighters required to transport the Dreamliner's components from around the world.

Airbus and the A380 Project

Airbus' economic geography is radically different to Boeing's and has its origins in the company's history. Airbus was originally a partnership formed between France's *Aerospatiale* and Germany's *Deutsche Airbus* to build the first twin-engine wide body aircraft (the A300). The Spanish company *CASA* joined the partnership followed in 1979 by *British Aerospace*. Each partner, known as Airbus France, Airbus Deutschland, Airbus UK and Airbus España, functioned as a national company with special responsibly for producing separate elements of the aircraft that were transported for final assembly in Toulouse. In 2001, this partnership of companies was collapsed into a single fully integrated company under the name *The European Aeronautic Defence and Space Company* (EADS); a merger of the French, German and Spanish interests holds 80% of the shares in the new company while BAE Systems, formerly British Aerospace, 20%. Like Boeing, Airbus draws upon the expertise and production facilities of a complex global supply chain. But, unlike Boeing, Airbus operates from 16 development and manufacturing plants that are positioned across Europe; production is organized via transnational manufacturing units that are responsible for producing a complete section of an aircraft for final assembly in Toulouse or Hamburg.

The original transnational partnership structure partially locks Airbus into a European production network. Politically, it would be difficult for Airbus to subcontract substantial elements of one of its products beyond Europe. This provides Boeing with a competitive advantage as its greater organizational flexibility enables it to develop partnerships with companies located in its key markets. Airbus is trying to follow a similar strategy. For the A380 project Airbus has entered into 200 major contracts with over 120 different suppliers and partners. These 'partners' have collaborated in the development and implementation of new technologies, working practices and design solutions. The company's supply chain consists of over 1,500 suppliers located in more than 30 countries. Airbus' difficulty is that within its core markets outside Europe supplier involvement in the A380 project is relatively limited. For example, in 2005 Gustav Humbert, Airbus chief executive, noted that 21 Japanese companies were already involved in the A380 programme, but only on a small scale, and he wanted greater Japanese involvement in the development of future aeroplanes. To establish a foothold in key foreign markets Airbus wants to outsource up to 60% of its production to countries such as China and India; China will be responsible for 5% and Russia 3% of the planned 253 seat A350 aircraft, Airbus' response to Boeing's Dreamliner (Jones 2005). Boeing may have been able to lock the major Japanese manufacturers into its supply chain. Boeing has long-established working relationships with companies like *Mitsubishi Heavy Industries* which is building the Dreamliner's wings and *Kawasaki Heavy Industries* that is responsible for some of the midfuselage sections; both companies are heavily involved with the Dreamliner's design. In October 2005, Airbus tried to persuade these companies to undertake contract work on the A350, but the Japanese companies refused to become involved due to order commitments to Boeing.

From this it is apparent that both Boeing and Airbus have and are in the process of establishing complex spatial divisions of labor and expertise. There are five drivers behind this evolving process: cost, access to expertise, sharing of development risk, access to markets and politics. Like Boeing, Airbus has developed a carefully structured and extremely complex spatial division of expertise. This strategy involves indirectly accessing expertise that has been developed to support Boeing's activities as well as developing a transnational, intrafirm spatial division of expertise. These are linked strategies. First, Airbus' supply chain is interwoven with that of Boeing;

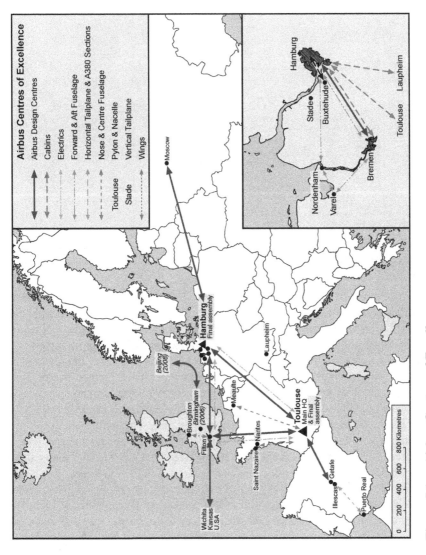

Figure 5.1　Airbus Centers of Excellence

Table 5.2 The spatial division of design expertise within Airbus

Country	Function
France	Toulouse Responsible for overall design, future projects, flight tests center and cabin design
United States	Wichita Wing design Opened 2002 (160 people)
United Kingdom	Filton Wing design, also future products, real time link to Wichita (USA) (2000 people) Birmingham Design office may be opened in 2006
Russia	Moscow Engineering Center Airbus, joint venture with Kaskol, Moscow, Design of cabin interiors, fuselage, supports engineers in Hamburg and Toulouse, opened 2003, (100 engineers)
Germany	Hamburg Product definition and overall aircraft design – including cabin design, systems engineering, structural design and testing for the fuselage, Future aircraft programmes team Bremen Design office – low speed aerodynamics, electric and hydraulic components and mobile parts
Spain	Getafe Design engineering of horizontal tail plane, rudder and rear sections, center for composites design and manufacturing
China	Beijing Technology transfer programme that will result in 200 engineers being employed at an Airbus engineering center by 2008, cooperation with AVIC I and AVIC II

they share many of the same suppliers. Over 100,000 American jobs are supported by Airbus contracts and around 40 per cent of the firm's global procurement budget is spent in the USA; around $5 billion per year on parts, components, tooling and services (Airbus 2005). In June 1986 Electroimpact, an independent company, was established in the shadow of Boeing's Everett plant to supply machine tools to the company. At the start of the 1990s the firm was a small Washington-based capital equipment supplier with 15 employees. By the mid 1990s the company was heavily involved with Airbus and is currently responsible for the supply of the four lines that assemble the A380 wings' large upper and lower panels. The company now has over 250 employees and has established subsidiaries to service its equipment in the UK and Israel. The British branch employs 20 engineers that provide 24/7 maintenance cover for Airbus' wing center of excellence at Broughton, North Wales.

Second, in 2004 Airbus reorganized the company by establishing six Centers of Excellence based around expertise in key production areas: wings at Filton and Broughton (UK); forward and aft fuselage at Nordenham, Varel, Bremen and Hamburg (Germany); nose and center fuselage at Toulouse, Saint Nazaire, Nantes and Méaulte (France); vertical tailplane in Stade (Germany; pylon and nacelle at Saint Eloi (France); horizontal tailplance and A380 sections at Getafe, Illescas and Puerto Real (Spain) (Figure 5.1). The Centers of Excellence are responsible for the design and development of aircraft components which are then assembled in Toulouse or Hamburg. This strategy encourages specialization as well as intrafirm, transnational cooperation. Like Boeing the strategy involves the development of a spatial division of labor as well as expertise. Airbus has established a network of design centers of excellence that enables it to capitalize on expertise that is situated within its core target markets. Seven national design offices have been established (Table 5.2). The majority of these are located within one of the core production centers of excellence. The Toulouse office is responsible for overall design while the other six offices specialize in individual elements of the aircraft. Like Boeing, the design offices are linked together in real-time.

In 2002, Airbus opened its first design and engineering center outside Europe in Wichita, Kansas. At Wichita, 160 engineers work on the A380 as well as specializing in wing design in tandem with the Filton office (UK). A location in Wichita enables Airbus to capitalize on the expertise that has been developed in this city by Boeing. It is worth noting that Spirit (Wichita) is responsible for the design and development of the Dreamliner's fixed and movable leading edges (wing components) (Table 5.1). In October 2005 Airbus announced that the Wichita office was recruiting an additional 70 engineers; 20 of these would work on the A350s wings including training edges and 50 on the design of cabin furnishing. In 2007, the cabin design team will transfer to an office that is being established in Mobile, Alabama (Vandruff 2005). In 2003, Airbus established a joint engineering center in Moscow with Kaskol, the Russian aeronautical company. The 100 engineers in this office are tied into project teams based in Hamburg and Toulouse. Airbus has also established a design office in Beijing with the Chinese aeronautical conglomerates *Aviation Industries of China I and II* (AVIC I and AVIC II). This is part of a Chinese knowledge transfer strategy with the intention being that this office will employ 200 engineers by 2008; currently the team is concentrating on design work for the A350.

The A380 project is surrounded by a complex global mosaic of companies that are providing both expertise as well as production capacity to support the project's development. These inputs can be classified as either direct (Airbus) or indirect (first- and second-tier suppliers) and as supporting (machine tools, logistics) versus being directly involved in the production of parts for the aircraft. The Airbus global international division of labor and expertise involves production as well as a broad range of service inputs including transport. Leif Høegh, the Oslo-shipping company, has ordered special cargo ships that will transport A380 components; Høegh Autoliners is a large company that transports new and used cars with a fleet of about 50 cargo ships. Together with its French partner, Luis Dreyfus Armateurs, Høegh Autoliners, was awarded the contract for all sea-born transport of components/parts for the giant aeroplane. Through a jointly owned French company, Dreyfus and Høegh, two new specially designed roll on/roll off cargo ships will be used to transport Airbus components for the next thirty years. The ships, are being built at the factory of 'Singapore Technology Marine' and will be in service by the end of 2008; each ship will cost around 30 million euros. This joint-venture already owns a ship, the 'Ville de Bordeaux', which sails for Airbus. The three vessels will sail between Hamburg, Cádiz (Spain) and St. Nazaire and Bordeaux in France. In addition to the aeroplane parts/components, the ships can also transport cars, trailers and other heavy, 'rolling' cargo. According to the administrative director of Høegh Autoliners, Thor Jørgen

Guttormsen, 'seen individually this does not represent a very big deal/engagement for us, but it is nice to be associated with such a prestigious project as the production of the world's largest aeroplane'. The Oslo-shipping company provides Airbus with expertise in brokering these types of specialist vessels and also played a central role in developing the ship design (Dagens Næringsliv, 2/2-2006).

Spatial Divisions of Expertise in Transnational Service Firms

The aerospace example explores the development of spatial divisions of expertise within complex manufacturing companies that are innovating by combining intra- and inter-firm spatial divisions of expertise. We must now consider the application of this concept to service firms. The production and consumption of services is traditionally considered as inseparable. This has important implications for the establishment of global or international service firms. Major accountancy practices can promote the firm's name and products, but the real marketing is through face-to-face interaction on the ground. Accountancy practices must create, develop and sustain relationships with a local client base (Greenwood et al. 1991). International service firms cannot concentrate their activities in global cities, but must develop a dispersed network of local offices, nationally and internationally. Many global professional service firms are confederations or co-operative alliances of independent national partnerships. For accountancy, this means that each major practice must have local offices or national partnership located in many different locations; each national partnership must be relatively self-sufficient and employ professionals that are embedded within the local society and culture as they must market and deliver services to a local client base.

    In Norway, a spatial division of expertise exists between local and national and in some cases international expertise that reflects the type and market demand for producer services (Bryson and Rusten 2005). There are a variety of ways in which a client located outside the city regions can access such expertise. They can access it by employing a consultant based, for example, in Oslo or Bergen and pay for the time and travel expenses required to obtain this type of expertise. There are problems with this technique as the Oslo consultant may be unaware of local business cultures. In many cases, expertise located in Oslo is accessed through an interaction that occurs between local and non-local experts. A good example is PwC Consulting which became part of IBM (International) in 2003. This firm operates in many different locations across Norway providing a combination of well-known business solutions and specialist expertise. The specialist expertise is provided by employees located in Oslo and accessed locally through the firm's network of local practitioners. This means that local clients experience specialists mediated through local experts who are aware of local business needs. Other local units provide more general services within accountancy and management advice. The relationship between local general expertise and non-local specialist expertise reflects not necessarily the advantages that accrue to global cities, but rather the market size: only some places can support the economic infrastructure (education support, number of local clients etc) required to maintain this level of expertise.

Local branch offices of the major producer service firms sometimes develop specializations or centers of excellence constructed around the needs of the regional business community in which they are embedded (Rusten et al. 2005). This is a very important point and one that is determined by the relationship that exists between the supply and consumption of expertise. Local demand will encourage the development of local specializations within knowledge intensive services. Branch offices can become so specialized that the global partnership identifies them as either a national or international center of excellence. This means that expertise which is developed to mirror the needs of a region's economic structure may become so important that the regional branch begins to provide sector and sometime functional expertise to other offices in the expertise provider's national and transnational network.

The global accountancy and consultancy firms have increasingly to transform themselves into deep specialists. This provides them with competitive advantage based around knowledge as well as reputation. Fenton and Pettigrew (2000) provide a detailed account of Coopers & Lybrand's strategy to develop global consultancy expertise in the pharmaceutical industry. At the time of the study, Coopers & Lybrand (C&L) employed 70,000 worldwide (30,000 in Europe) in around 140 countries. Whilst C&L is a global service firm only between 15-20 per cent of its consulting business was transnational, either working across borders or in cross-border teams. Only around 10 per cent of staff worked regularly outside their domestic environment. This highlights the local nature of much service work and places global service provision in a wider context. However, all employees share the firm's online resource base and this creates possibilities for the exchange of ideas as well as enabling professionals to join forces 'behind the front desk'.

C&L is a federation of small autonomous partnerships. In 1989, in order to integrate the autonomous partnerships, Coopers & Lybrand Europe was established. CLE was an integrated network of national partner firms based in 16 countries. This was a virtual firm as it had no head office but London and Brussels was its hub. CLE was C&L first regional unit. Prior to the establishment of CLE the firm had no mechanism to manage global activities. Management Consultancy Services (MCS) was developed as an off-shoot of CLE as it was considered that management consultancy needed to develop to respond to global forces. Strategic decisions could now be taken globally and it was decided to co-ordinate team plans in Europe for the development of four industry programmes (retail and consumer, pharmaceuticals, telecoms and financial services). This would provide MCS with deep knowledge about these sectors. A spatial division of expertise was developed via the creation of sub-networks between the national firms that produced cross-fertilization.

The pharmaceutical sector consists of a small number of clients that operate in multiple countries. MCS worked with clients through a partner or director who had contractual responsibilities to commit the firm to the project. National firms supplied the consultants for temporary project teams. The national firms had from one to 31 pharmaceutical specialists with the largest groupings in London and Uxbridge

(31), Basel (14), Utrecht (11), Brussels (11), Milan (8) and Frankfurt (7). A project manager was appointed with day-to-day responsibilities for managing the project and for very large projects team leaders were appointed under the control of the project manager. Consultancy was delivered through two levels of consultants: experienced professionals supported by less experienced junior consultants. MCS developed an integrated European business plan to develop its pharmaceutical business. It established a virtual management team with members located in different regional offices across Europe. The *Pharma Network* consisted of around 200 consultants from across Europe and between 1996 and 1998 its business increased by 30-40 per cent per year (from 20 to 45 million pounds). MCS has a major problem. It tries to act like a European firm providing global expertise, but a national mentality still dominates. This means that resources could not be utilized effectively as some national firms were able to influence the network's operation. The Pharma Network works beyond global cities by unlocking potential expertise synergies that exist within MCS. Clients may access the network from London, but obtain expertise that is located and has been developed in other places. In the same way, clients can access the network from any access node (place) and may never obtain expertise that is placed within the global city. Like Boeing and Airbus, the MCS spatial division of expertise has been developed to support local clients and the place of its production is related to the geography of the pharmaceutical industry.

The Pharma Network was led by a Chairman who was a UK partner dedicated to the task and no longer attached to a national firm. The Pharma Network's success helped to develop a common culture; shared values, a corporate language and a framework for understanding client's business and their needs developed. Common values and beliefs began to erase the possibilities for opportunistic behaviour by individual consultants who began to consider themselves to be part of the network rather than representing their national partnership. The network developed informal meetings as well as organising a *Pharma University* to share knowledge and expertise. The formal and informal contacts were informed by a database of phamaceutical consultants' CVs and each consultant was provided with the names and contact details of all pharma consultants.

Fundamental to the Pharma Network was the use of reputation as a form of social control. Reputation, or knowledge about the reliability and skills of individual consultants, diffuses across the network. Thus, 'the frequent exchanges and recombinations of consultants on projects meant that information about individuals and firms travelled quickly. A branch unwilling to co-operate, either in their interface with the client or with the network, would be exposed to collective sanctions by other firms' (Fenton and Pettigrew 2000, 102). Firms that had acquired a good reputation amongst network members were asked to join more multinational teams.

The example of Coopers & Lybrand highlights the importance of individuals and their reputations in the management and delivery of knowledge-intensive services. It also highlights some of the distinctive features of expertise labor; client firms establish long-term relationships with individual consultants and frequently clients will follow consultants as they move between firms. It also substantiates some of the problems faced by firms trying to deliver global products locally. The delivery of knowledge-intensive service rests on the orchestration of networks – intra-firm networks, with clients and with other service suppliers. The position of a firm within a network is both enabling and constraining. It provides opportunities to access clients and resources, but also is a constraint as a firm and an individual may experience lock-in; the costs of breaking from the network or switching to another network may be too high.

Conclusion

Change is a fundamental feature of capitalism as individuals and firms attempt to maximise profit by responding to alterations in consumption as well as production. New technologies open opportunities for new forms of production, organization and work. The enhanced importance of expertise is central to the current on-going transformation of capitalism. This expertise takes many different forms including expert consumers as well as the growth of in-house as well as external producer services. The existing academic literature has focussed on understanding the dynamics of producer service firms located in global cities. The starting and in many cases the end point of the analysis has been on this handful of cities. The starting point or initial set of questions of a research project influences and in many cases determines the outcome. We argue, along with Massey (1984), that the primary focus of economic geography should be on understanding the spatial organization of the social relations of all types of manufacturing and service production. Given this the starting point of the analysis should not be on the global cities but on the spatial organization of production. Ultimately, this might mean that the research would eventually highlight the important role global cities play in the organization of the global economy, but other 'global' places and processes would also be made visible.

Our argument is complex as it attempts to link two distinctive and on-going debates. We begin with a critique of the over-emphasis placed on global cities while the end point has been a call for the development of an academic project that explores the development, organization and operation of 'spatial divisions of expertise'. The latter acknowledges that expertise-intensive work is distinctively different from that conventionally associated with the term labor. In occupations or professions constructed upon capitalising on individuals' expertise and reputations much of a firm's value is contained in people rather than in capital and equipment. This transforms the employment relationship as a firm's primary competitive advantage is highly mobile. The embodied nature of producer service work transfers the power in the employment relationship away from the employer to the employed.

The enhanced importance of expertise and the rapidity with which demands for specific types of expertise can develop have increased the importance of the spatial division of expertise. Expertise labor is highly mobile and is able to relocate to satisfy local demands. The global economy is experiencing increasing international labor mobility. Geographical variations in skilled labor, new and unanticipated demands for specialist expertise and the operation of national educational systems are important drivers behind this enhanced mobility. Norway is currently experiencing an important engineering boom as the country's petroleum industry develops oil and gas resources and related inland and export pipeline networks. This boom has created a demand for an additional 10,000 engineers. As a result, highly skilled engineers are being recruited from several countries to work in Norway or elements of project engineering terms have been offshored (Rusten et al. 2006). This Norwegian example describes the on-going development of a complex spatial division of expertise that binds experts located in many different places together into project teams.

We have shown through the aerospace and Pharma Network examples that different types of expertise have different places or spaces of formation or production. The global cities are associated with advanced financial services and related producer services, but they are not the locations for all types of expertise. In many cases, expertise is accessed via the global cities that is located in other places but linked, for example, to London and New York by formal or informal networks. Economic geographers need to identify and unravel these networks as these spatial divisions of expertise represent the visible as well as invisible architecture of national and international economies.

References

Airbus (2005), *Airbus in North America*, www.airbus.com/en/worldwide/airbus_in_ north-america.html, accessed 26/09/2005.

Beaverstock, J.V. (2004), 'Managing across Borders: Transnational Knowledge Management and Expatriation in Legal Firms', *Journal of Economic Geography* 4, 157–179.

Braverman, H. (1974), *Labor and Monopoly Capital: The Degradation of Work in the Twentieth Century* (New York: Monthly Review Press).

Bryson, J.R. (1997), 'Business Service Firms, Service Space and the Management of Change', *Entrepreneurship and Regional Development* 9, 93–111.

Bryson J.R. and Daniels, P.W. (2006), 'A Segmentation Approach to Understanding Business and Professional Services in City-Regions: Shifting the Horizon beyond Global Cities', in Rubalcaba, L., Kok, H. and Baker, P. (eds), *Business Services in European Economic Growth* (Basingstoke, Hampshire: Palgrave Macmillan).

Bryson, J.R. and Monnoyer, M.C. (2004), 'Understanding the Relationship between Services and Innovation: The RESER Review of the European Service Literature on Innovation', *The Service Industries Journal* 24: 1, 205–222.

Bryson, J.R. and Rusten, G. (2005), 'Spatial Divisions of Expertise: Knowledge Intensive Business Service Firms and Regional Development in Norway', *The Services Industries Journal* 25: 8, 959–977.

Bryson, J.R., Keeble, D. and Wood, P. (1993a), 'Business Networks, Small Firm Flexibility and Regional Development in UK Business Services', *Entrepreneurship and Regional Development* 5: 3, 265–277.

Castells, M. (1996), *The Information Age: Economy, Society and Culture, Vol 1: The*

Rise of the Network Society (Oxford: Blackwell).

Christaller, W. [1933] (1966), *Central Places in Southern Germany*, translated by Baskin, C.W. (Englewood Cliffs, N.J.: Prentice-Hall).

Clark, D. (1996), *Urban World/Global City* (London: Routledge).

Daniels, P.W. (1991), 'Some Perspectives on the Geography of Services', *Progress in Human Geography*, 15: 1, 37–47.

Daniels, P.W. and Bryson, J.R. (2002), 'Manufacturing Services and Servicing Manufacturing: Changing Forms of Production in Advanced Capitalist Economies', *Urban Studies* 39: 5–6, 977–991.

Daniels, P.W. and Bryson, J.R. (2005), 'Sustaining Business and Professional Services in a Second City Region: The Case of Birmingham, UK', *The Service Industries Journal* 25: 4, 505–524.

Dicken, P. (2003), *Global Shift: Reshaping the Global Economic Map in the 21ˢᵗ Century*, (London: Guilford).

Doh, J.P. (2005), 'Offshore Outsourcing: Implications for International Business and Strategic Management Theory and Practice', *Journal of Management Studies* 42: 3, 695–704.

Durkheim, E. (1933), *The Division of Labour in Society* (New York: Macmillan).

Fenton, E.M. and Pettigrew, A.M. (2000), 'The Role of Social Mechanisms in an Emerging Network: the Case of the Pharmaceutical Network in Coopers & Lybrand Europe', in Pettigrew, A.M. and Fenton, E.M. (eds), *The Innovating Organization* (London: Sage), 82–116.

Florida, R. (2002a), *The Rise of the Creative Class and How It's Transforming Work, Leisure, Community and Everyday Life* (New York: Basic Books).

Florida, R. (2002b), 'The Economic Geography of Talent', *Annals of the Association of American Geographers* 92: 4, 743–755.

Gapper, J. (2006), 'Only the brave can make jet engines', *Financial Times*, February 20, 17.

Gates, D. (2005a), 'Boeing 787: Parts from around world will be swiftly integrated', *The Seattle Times*, September 11.

Gates, D. (2005b), '787 work advances on multiple fronts', *The Seattle Times*, Business and Technology, September 11.

Goldthorpe, J. (1982), 'On the Service Class, its Formation and Future', in Giddens, A. and Mackenzie, G. (eds), *Social Class and the Division of Labour* (Cambridge: Cambridge University Press), 162–85.

Goldthorpe, J. (1995), 'The Service Class Revisited', in Butler, T. and Savage, M. (eds), *Social Change and the Middle Classes* (London: UCL Press), 313–29.

Greenwood, R., Rose, T., Brown, J.L., Cooper, D.J. and Hinings, B. (1999), 'The Global Management of Professional Services: the example of accounting', in Clegg, S., Ibarra-Colado, E. and Bueno-Rodriquez, L. (eds), *Global Management: Universal Theories and Local Realities*, Sage: London, 265–297.

Hanlon, G. (1994), *The Commercialisation of Accountancy – Flexible Accumulation and the Transformation of the Service Class* (London: Macmillan).

Hanlon, G. (1999), 'International Professional Labour Markets and the Narratives of Accountants', *Critical Perspectives on Accounting* 10, 199–221.

Hoddos, J. (2002), 'Globalisation, Regionalism and Urban Restructuring: The Case of Philadelphia', *Urban Affaires Quarterly* 37, 358–379.

Jones, D. (2005), 'Airbus Work to be Sent to Japan', *Daily Post*, October 27.

Kageyama, Y. (2005), 'Japanese Manufacturers Rebuff Airbus Overtures', *The Seattle Times*, October 27.

Knight, G.A. and Cavusgil, S.T. (2004), 'Innovation, Organizational Capabilities, and the Born-Global Firm', *Journal of International Business Studies* 35: 2, 124–42.

Markusen, A.R., Lee, Y.-S. and DiGiovanna, S. (eds) (1999), *Second Tier Cities: Rapid Growth Beyond the Metropolis* (Minneapolis: University of Minnesota Press).

Marshall, J.N. and Wood, P. (1995), *Services and Space* (Harlow: Longman).

Marx, K. [1844] (1981), 'Economic and Philosophical Manuscripts', in Marx, K., *Early Writings* (Harmondsworth: Penguin).

Massey, D. (1984), *Spatial Division of Labour: Social Structures and the Geography of Production* (London: Macmillan).

MacMillan, A. (2005), 'Smooth Landing for LFM', LFM-SDM News, http://lfmsdm.mit.edu/news_articles, accessed 9/11/2005.

O'Farrell, P.N., Moffat, L.A.R. and Hitchens, D.M.W.N. (1993), 'Manufacturing Demand For Business Services in a Core and Peripheral Region: Does Flexible Production Imply Vertical Disintegration of Business Services?' *Regional Studies* 27, 385–40.

Rusten, G. (2000), 'Geography of Outsourcing: Business Service Provisions Among Firms in Norway', *Journal of Economic and Social Geography TESG* 91: 2, 122–134.

Rusten, G., Bryson, J.R. and Gammelsæter, H. (2005), 'Dislocated versus local business service expertise and knowledge: the acquisition of external management consultancy expertise by small and medium sized enterprises in Norway', *Geoforum*, 36. 525-539.

Rusten, G., Bryson, J.R. and Stabell, M.C. (2006), 'Offshoring and global sourcing by Norwegian organisations', paper presented at annual conference of the Association of American Geographers, Chicago.

Rusten, G., Gammelsæter, H. and Bryson, J.R. (2004), 'Combinational and Dislocated Knowledge and the Norwegian Client-Consultant Relationship', *The Service Industries Journal* 24: 1, 155–170.

Sayer, A. and Walker, R. (1992), *The New Social Economy: Reworking the Division of Labour* (Oxford: Blackwell).

Sassen, S. (2002), 'Locating Cities on Global Circuits', *Environment and Urbanization* 14: 1, 13–30.

Scott, A.J. (2002), *Global City Regions: Trends, Theory, Policy* (Oxford: Oxford University Press).

Smith, A. [1776] (1977), *The Wealth of Nations* (Middlesex: Pelican).

Starbuck, W.H. (1992), 'Learning by Knowledge-Intensive Firms', *Journal of Management Studies* 29: 6, 713–740.

Taylor, P. and Walker, D. (2001), 'World Cities: A First Multivariate Analysis of Their Service Complexes', *Urban Studies* 38: 1, 23–48.

Warf, B. (2000), 'New York: the Big Apple in the 1990s', *Geoforum* 31, 487–499.

Vanduff, K. (2005), 'Airbus Has Design on SKT Office Space', *Wichita Business Journal*, October 16.

Chapter 6

Internationalization of Management Consultancy Services: Conceptual Issues Concerning the Cross-Border Delivery of Knowledge Intensive Services

Joanne Roberts

The internationalization of services has attracted much attention in recent years. This interest began in the 1980s and increased with the inclusion of services on the agenda of the Uruguay round of the General Agreement on Tariffs and Trade (GATT) negotiations that began in 1986 (see for example: Shelp 1981; Riddle 1986; Nusbaumer 1987; Nicolaides 1989; *inter alia*). It further developed through the 1990s, and since the conclusion of the General Agreement on Trade in Services (GATS) in 1993, the internationalization of services has become a key issue of interest in the global economy (Daniels 1992; Aharoni and Nachum 2000; Miozzo and Miles 2002; Cuadrado-Roura, Rubalcaba-Bermejo and Bryson 2002; *inter alia.*).

The internationalization of producer services, and business services more specifically, has attracted the interest of many researchers (Aharoni 1993; Roberts 1998, Noyelle and Dutka 1988; O'Farrell et al. 1995; *inter alia.*), as has the internationalization of certain sub-sectors, such as advertising (Nachum 1999; Leslie 1996; Mattelart 1992; Perry 1990), accountancy (Daniels et al. 1989, Aharoni 1999), computer services (OECD 1989; OECD 1990; Kozul-Wright and Howells 2002) and more recently legal services (Beaverstock, Smith and Taylor 1999; Spar 1997) and the temporary staffing industry (Ward 2001; 2004). Although, the internationalization of management consultancy has attracted some attention (Kipping and Sauviat 1996; Jones 2003; UNCTAD 2002) the cross-border delivery of these services remain poorly appreciated.

Many services, especially business services, are knowledge intensive. Knowledge-intensive Business Services (KIBS), which are concerned with the collection, analysis and distribution of information and knowledge, play a significant role in the creation, dissemination and application of knowledge both within and between firms at the level of the region, the nation and internationally (Miles et al. 1995; Antonelli 1999; Andersen et al. 2000). This chapter therefore focused on management consultancy services, as an example of KIBS, with a view to identifying the issues for internationalization that face this sub-sector, and particularly the conceptual issues concerning the cross-border delivery of knowledge intensive services. Moreover, a conceptual framework is proposed which analyses the internationalization of management consultancy services in terms of the cross-border transfer of knowledge.

The chapter begins with the definition of management consultancy services and an account of their growth and development. This is followed by a discussion of the internationalization of UK management consultancy services before a brief review of the cross-border transfer of knowledge is presented. The role of management consultancies in the cross-border transfer of knowledge is then considered. A framework that seeks to capture the essence of the cross-border knowledge transfer facilitated through management consultancy services is then proposed. Building on this, the chapter concludes with an agenda for further research.

The Nature and Development of Management Consultancy Services

The growth of management consultancy is closely linked to the evolution of management practice and ideology (Kipping 2002). For instance, the primary aim of Arthur D. Little Inc., established in 1886, was to develop new technical solutions for the problems experienced by the growing sector of industrial enterprises in the United States (UNCTAD 2002, 4). The emergence and diffusion of scientific management in the early 20th century marked the formal development of management consultancy activity. Initially a range of actors provided management consultancy services from bankers to engineers (Ferguson 2001). However, the rapid growth of the sector did not begin until the Great Depression when legislative changes, including the Glass-Steagall Banking Act of 1933 in the US, which prohibited commercial banks from engaging in 'non-banking activities' like management engineering, promoted the rapid growth of independent management consultants (McKenna 1995).

Greiner and Metzger (1983, 7) define management consulting as:

> an advisory service contracted for and provided to organizations by specially trained and qualified persons who assist, in an objective and independent manner, the client organization to identify management problems, analyze such problems, recommend solutions to these problems, and help, when requested, in the implementation of solutions.

This is a broad definition that captures the key characteristics of management consulting. However, providing a more specific definition of management consulting is difficult give the continuously changing nature of the activity (Clark and Fincham 2002). Indeed, the services supplied by management consulting firms are subject to changing fashions in the field of management (Abrahamson 1996; Kieser 1997). Kipping (2002, 38) identifies three waves in the evolution of management consultancy, which he associates with the changing fortunes of the top consultancy firms. The first wave of scientific management was dominant from the 1930s–1950s, followed by the organization and strategy wave, which dominated from the 1960s – 1980s, and, thirdly, IT network building consultancies, dominant since the 1990s. To

Table 6.1 UK management consultancies association members' fee income by service line in 2003

Service line	Fee Income £ Million	Percentage of Total Fees*
Outsourcing-related Consulting	2,108	36.3
IT-related Consulting	1,374	23.6
Programme/Project Management	703	12.1
Human Resources	543	9.3
Strategy	497	8.6
Operations	155	2.7
Business Process Re-engineering	120	2.1
Financial	107	1.8
Change Management	88	1.5
Marketing & Corp. Communications	79	1.4
E-business	21	0.4
Economic and environmental	16	0.3
Total	**5,811**	**100.0**

* Figures rounded to the first decimals point.

Source: compiled and calculated from data presented in Czerniawska (2004, p. 23).

these three waves can be added outsourcing, which is currently the most significant consulting activity.

In terms of industrial classifications management consultancy falls under division K – 'Real estate, renting and business activities' – of the Eurostat NACE Rev 1. Specifically, within division 74 'Other business services', group 74.1 which includes: legal, accounting, book-keeping and auditing activities; tax consultancy; market research and public opinion polling; business and management consultancy; and, holdings. IT related consulting, which many management consultancy firms provide, falls under division 72 'Computer and related activities'. Business services have become increasingly recognized as of considerable significance to the competitiveness of enterprises generally and an important factor driving long-term growth (OECD 1999; CEC 1998; 2001). The European business service sector has experienced significant growth in the past 20 years. Since the mid-1980s employment in the sector has grown at a rate of approximately five per cent per annum, reaching a total employment level of over 12 million by the late 1990s (CEC 2000, 507). As illustrated later, management consultancy is an important contributor to these trends in 'Other business services'.

Many specialist branches of management consultancy exist today. These sub-groups are variously defined by different organizations depending on their specific interests. The UK Management Consultancies Association (MCA), whose members account for approximately 60 percent of the UK market, identifies the service lines detailed in Table 6.1. It is evident the outsourcing-related consulting and IT-related consulting account for the largest portions of MCA member fee income with £2,108 million and £1,374 million respectively (Czerniawska 2004, 23).

A major difficulty when estimating the size of the market for management consultancy services is that the definition of management consultancy varies and hence measures vary accordingly. Secondly, management consultancy services are delivered not only by management consulting firms but also by a range of business service providers in related areas, including, perhaps most significantly, accountancy firms, as well as advertising groups, public relations and market research companies and engineering consultancy firms (UNCTAD 2002). Thirdly, there is the difficulty of assessing the contribution of in-house consulting divisions, or corporate affiliates, of major manufacturing or service corporations. Finally, many firms in the sector are organized as partnerships, and, consequently do not have to disclose the exact amount of their revenues to the public. It is then difficult to define with any precision the value of the management consultancy sector.

Bearing these reservations in mind, the European Federation of Management Consulting Associations (FEACO 2003, 4) estimate the European management consultancy market size to be € 46.5 billion in 2002 (down from € 47.5 billion in 2001). Given that the US traditionally accounts for half of the world market, with Europe following closely behind (UNCTAD 2002), the world market can be estimated at well over $90 billion in 2002. Indeed, Kennedy Information (2004, 1) predict that even with slow growth the global market will be worth just under $125 billion in 2004.

According to a FEACO survey (2003, 4), there were approximately 57,000 enterprises with 300,000 consultants operating in the European market in 2002, suggesting that the majority of firms are small in terms of employment. Small sized firms and medium sized firms accounted for 8.7 per cent and 38.3 per cent of the market respectively. While the top 20 firms accounted for 53 per cent of the market in 2002, an increase from 47.5per cent in 1999, indicating a trend towards concentration in the market (FEACO 2003, 7). This trend is reflected in the UK market, where of the top 75 firms the first 10 accounted for 74 per cent of their total fee income in 2003 (Table 6.2), compared to 26 per cent for the remaining 65 firms (Management Consultancy 2004, 5).

For the first time in more than 25 years the European market for management consultancy services contracted in 2002. Between 1999 and 2001 the market grew by an annual average rate of 14.5 per cent but in 2002 it decreased by 2 per cent (FEACO 2003, 5). This trend was not evenly distributed across the European market, with markets such as the UK's continuing to grow but at slower rate. The

Table 6.2 Top 20 UK management consultancy firms ranked by fee income F/Y 2003

Rank	Firm	Fee income £m 2002	Fee income £m 2003	Financial Year End
1	Accenture	565.0e	573.0e	31.12.03
2	IBM	490.0e	500.0e	31.12.03
3	AtosOrigin	409.0e	417.0e	31.12.03
4	Xansa plc	411.3e	410.0e	30.04.04
5	Deloitte	336.0e	342.0e	31.05.04
6	Capgemini	332.0e	250.0e	31.12.03
7	PA Consulting Group	174.7	183.0	31.12.03
8	LogicaCMG	162.0e	170.0e	31.12.03
9	Capita	145.0e	160.0e	31.12.03
10	McKinsey & Co.	147.0e	151.0e	31.12.03
11	Fujitsu	128.0e	130.0e	31.12.03
12	Towers Perrin	89.0	92.0	31.12.03
13	CSC Computer Sciences Corp	80.0e	58.6e	31.12.03
14	Bain & Company	54.0e	56.0e	31.12.03
15	Detica	39.2	50.0e	31.03.04
16	Boston Consulting Group	43.0e	46.0e	31.12.03
17	Penna Consulting plc	47.0	42.0e	31 .03.04
18	BDO Stoy Hayward	38.0e	40.0e	30.06.03
19	HEDRA	30.5	37.5	31.03.04
20	BT Syntegra	34.0e	36.0e	05.04.04

e = estimate. All estimates are those of *Management Consultancy*.

Source: Adapted from Management Consultancy, May 2004, p. 5. (Also available at: http://www. managementconsultancy.co.uk/bif_static/pdf/MC_top25.gif last accessed 16th June 2004).

European market now contributes 0.42 per cent of the European GDP well above the contribution to global output estimated to be 0.25 per cent, and an increase of 250 per cent on the 1994 figure (FEACO 2003, 6). The contribution to GDP varies considerably across Europe, with the UK sector contributing 1 per cent to GDP, with the second largest contributor being Germany with 0.48 per cent and Greece's sector contributing only 0.12 per cent to GDP. Clearly there is much scope for the growth of the market throughout Europe and especially in the new members of the EU. The UK and Germany are the largest European markets for management consultancy with 29.2 per cent and 28.0 per cent share of the European market respectively, followed by France with 13.4 per cent (FEACO 2003, 11). The UK has the highest turnover per consultant with € 270,000, followed by Switzerland with € 230,000 and Germany with € 190,000 (FEACO 2003, 12).

Internationalization of UK Management Consultancy Services

A number of studies have explored the internationalization of management consultancy services (Noyelle and Dutka 1988; UNCTAD 2002; Jones 2003). At a broad level the internationalization of management consultancy can be linked to the diffusion of new management practices, in particular the diffusion of US management practices and consulting methods to Western Europe, especially after World War II (Kipping 1996; Bryson 2002). However, here the concern is with the internationalization of management consultancy firms and how these organizations facilitate the cross-border transfer of knowledge.

The ongoing process of the globalization of economic activity is a key force promoting the internationalization of management consulting firms. As client firms become increasingly international, their suppliers of business services including management consultancy services follow them into foreign markets by establishing an overseas affiliate (Roberts 1998). However, UNCTAD's (2002) analysis of the tradability of management consultancy products suggests that such patterns may be changing. Indeed, the UK exports of Business Management and Management Consultancy services increased from £491 million in 1992 to

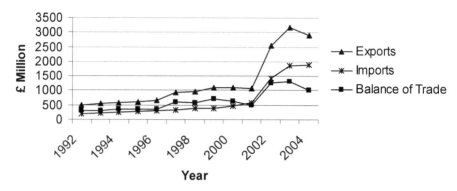

Figure 6.1   UK Trade in business management and management consulting services

Source: Compiled from data available in Table 3.9 of the UK Balance of Payments Pink Book 2003 and 2005 edition.

£2,899 million in 2004, over the same period imports rose from £184 million to £1,881 million (Figure 6.1). A significant increase in UK trade in these services occurred between 2001 and 2002, perhaps resulting from the continued growth of the UK sector alongside a general down turn in the European market. The UK market may have absorbed some of the excess capacity from other European countries in terms of imports, while UK consultants would have had a stronger domestic position upon which to build their own exports.

During the 1990s the UK had a persistent surplus in the balance of trade in this area. Moreover, in 2002 Business Management and Management Consultancy services accounted for approximately 10 per cent of all exports and imports in the 'Other business services' category.

Nevertheless, it is important to note that the value of trade in these services is relatively low compared to the domestic value of the UK market, which is estimated to have been worth £10 billion in 2003 (Czerniawska 2004, 11). Czerniawska (2004, 14) estimates that in 2003 non-UK based clients generated 10 per cent of fees in the UK, down from 14 per cent in 1995 (Table 6.3). Czerniawska suggests that part of this change may be attributed to changes in the way firms account for overseas income internally. Furthermore, UK firms find it hard to expand abroad partly because of relatively high UK fee rates and partly as a result of the depressed markets in the Euro-zone. Finally the development of continental consulting firms may be an additional factor resulting in overseas clients being supplied in their home markets rather than from the UK.

Table 6.3 Geographical source of fee income for UK management consultancies association members 1995 and 2003

Location	1995		2003	
	£ Million	Percentage	£ Million	Percentage
UK	1,084	86	5,248	90
EU excluding UK	82	7	248	4
Rest of the World	90	7	315	6
Total	1,256	100	5,811	100

Source: Compiled from data presented in Czerniawska (2004, pp. 13–14).

Methods of Internationalization Among Management Consultancy Service Providers

Many classifications of the various methods of cross-border delivery of service have been developed (Sampson and Snape 1985; UNCTAD 1983; Stern and Hoekman

Table 6.4 Modes of cross-border service supply set out in GATS

Mode	Description
1	Cross-border supply of services corresponds with the normal form of trade in goods.
2	Consumption abroad, or in the words of Article I the supply of a service in the territory of one Member to the service consumer of another Member. Typically, this will involve the consumer travelling to the supplying country, perhaps for tourism or to attend an educational establishment.
3	The supply of a service through the commercial presence of the foreign supplier in the territory of another WTO member. Examples would be the establishment of branch offices or agencies to deliver such services as banking, legal advice or communications.
4	The presence of natural persons, or, the admission of foreign nationals to another country to provide services there.

Source: Compiled from WTO (1999).

1988; *inter alia.*)[1] However, currently the most influential classification of modes of cross-border service delivery is that detailed in Article I of GATS in which a comprehensive definition of trade in services is set out in terms of four different modes of supply: cross-border, consumption abroad, commercial presence in the consuming country and presence of natural persons (Table 6.4).

These modes of cross-border service delivery are not mutually exclusive. A firm may use several simultaneously, for example, the movement of personnel may accompany cross-border trade. The method used to deliver a service will depend on a variety of factors including the nature of a service. Those management consultancy services that can be standardized on an international basis can be produced internationally. Whereas those services that require specific knowledge of the national base are more dependent on local production whether by local firms or subsidiaries of international firms. (UNCTAD 2002, pp. 51–52).

It is also important to note that the cross-border delivery of services within business services may actually take place within the boundaries of the firm in the form of intra-firm trade (Roberts 1998, 1999). For instance, manuals may be distributed, or databases accessed, throughout the global network constituting intra-firm trade, yet to facilitate the cross-border delivery of a service to a client firm

---

1   See Bryson, Daniels and Warf (2004) for a review of the classification of international service transactions.

requires that the knowledge embodied in the manual be extracted and applied by a consultant *in-situ*.

Each of the four modes of cross-border service delivery gives rise to specific types of barriers with the subsequent restriction on the globalization of the market for services. In addition to barriers to the cross-border supply of service trade in the form of tariffs and subsidies, services are also affected by regulations on the mobility of customers and personnel as well as on flows of foreign direct investment, and particularly the right of establishment. In some business service sectors there are specific regulatory barriers to internationalization. For instance, in the legal and accountancy sectors familiarity and qualifications in national practices may be required. Although there are generally no specific regulatory barriers to the cross-border supply of management consultancy services, restrictions on the movement of personnel and clients can limit access to certain markets. In addition, barriers to entry into new markets do exist in terms of access to networks (Glückler and Armbrüster 2003).

UNCTAD (2002) provides an assessment of the tradability of a number of categories of management consulting services (Table 6.5). According to this assessment activities related to finance and IT consulting, and economic studies demonstrate a good level of tradability, whereas the tradability of other activities like human resource consultancy are poor.

The Cross-Border Transfer of Knowledge

In the international context it is possible to identify a number of channels facilitating the cross-border transfer of knowledge (Howells and Roberts 2000). Firstly, Multinational Enterprises (MNEs) have a significant role not only in the creation of new knowledge and technology in an international environment but also as facilitators of international interactions of knowledge both within the boundaries of the firm and externally through trade in goods and services, and the sale of technology through licensing agreements. Secondly, international technical alliances give rise to knowledge interactions between firms from different countries, and between MNEs and national based firms. Thirdly, knowledge interactions occur through the international trade of capital, intermediate and final goods. Fourthly, international interactions of knowledge are facilitated through mobile labour, in particular managerial, and science and technology personnel. For instance, knowledge transfer is facilitated through cross border collaboration between researchers internal to the firm or external through joint international science projects. Fifthly, trade and foreign direct investment in services, such as technical and management consultancy services facilitate knowledge interactions. Finally, knowledge interaction occurs through a wide variety of social and cultural cross-border mechanisms, including those facilitated by the media and information and communication technologies (ICTs) more generally.

Table 6.5 Management consultancy services and their tradability

Product Categories	Specific Services	Tradability
General management	Diagnostic survey	Some
	Corporate strategy	Some
	Structures and systems	Some
	Corporate culture and management style	Poor
	Innovation and entrepreneurship	Poor
Financial and administrative systems	Financial appraisal	Good
	Working capital and liquidity management	Good
	Capital structure and financial markets	Good
	Mergers and acquisitions	Fair
	Capital investment analysis	Good
	Accounting systems and budgetary control	Fair
Marketing consultancy	Marketing strategy	Fair
	Implementation of marketing activities	Some
	Market research	Good
Production and services management	Product design	Good
	Production and organisation	Poor
	Quality control	Poor
Human resources	Human resource planning	Poor
	Recruitment and selection	Poor
	Human resource development	Poor
Information technology and systems	Advice on purchase of software and hardware	Good
	Adaptation of software products	Fair
	Development of new software	Good
	Outsourcing	Good
Economic studies	Economic studies	Good
Ecology and environmental issues	Waste management and pollution	Good
	Regulatory issues	Poor
	Working environment and safety	Poor
Project management	Project management	Fair

Source: Adapted from UNCTAD (2002, p. 111).

In a study of international technology transfer in services Grosse (1996, 790) identifies the following means of technology transfer: hardware (machinery), software, people transfer, people training, documentation, communication, and, agreements (permissions). Results from the survey of Latin American service firm affiliates indicate that these means are not used in a mutually exclusive manner. Methods used in descending order of importance were: training programmes, manuals, visits by experts, and employment of expatriates (Grosse 1996). In addition, Grosse includes among the vehicles for technology transfer foreign direct investment, licensing, technical assistance contract, training contract, turnkey contract, representation contract, exporting, franchising, management contract, R&D contract, co-production agreement, and, subcontracting.

Indicators of international knowledge interactions are well advanced at a general level. They include data on the flows of technology payments, global diffusion of patents, trade in embodied technology and joint R&D consortia. According to the OECD (1997, 29) these indicators are increasing over time for all OECD countries, although at different levels and rates, indicating the growing significance of international knowledge transfer. However, these indicators fail to capture all international knowledge transactions. In addition to the flows listed above, knowledge diffused by international organizations such as the World Bank and the OECD, as well as non-governmental organizations, like Greenpeace, should also be included (Howells and Roberts 2000). Moreover, existing measures reflect transfers of codified or explicit knowledge, which can be embodied in a patent or contract, where as the transfer of tacit knowledge is more difficult to assess since it is transferred through a process of learning by doing (Polanyi 1958). The successful cross-border transfer of tacit knowledge requires the movement of people and the development of relationship of trust and mutual understanding between the producer and the client (Roberts 2000). Consequently, measuring the transfer of tacit knowledge is problematic.

The Role of Management Consultancies in the Cross-Border Transfer of Knowledge

The mechanisms through which management consultancy services are delivered across borders provide the channels through which knowledge is transferred. For if a service is to be provided this requires a process of knowledge transfer, firstly from the client firm to the provider and secondly from the provider to the client firm. In addition, the international delivery of services may require the transfer of knowledge between a range of consultancy affiliate firms in a number of countries as well as inputs from a number of client affiliates within a global network. Intra-firm trade may well be essential in facilitating the production and delivery of the final service.

However, different internationalization mechanisms favour specific forms of knowledge transfer. Indeed, a number of mechanisms may be utilizes in the delivery of a service. For instance, a movement of personnel to deliver a report may include a presentation and face-to-face contact as well as cross-border trade such as the delivery of a report in a tangible or electronic form.

Grosse's (1996) empirical research suggests that the most important internationally technology transferred in the services sectors studied was some kind of personal knowledge held by employees. For the management consulting firms included in Grosse's research the most important forms of knowledge transferred related to experience in the business, methodology for producing the service and technical information.

But how is knowledge transferred between locations in a global management consultancy firm and between the firm and clients? In the following subsections key mechanisms through which knowledge is transferred are considered.

A Global Corporate Culture

The development of a global corporate culture is a vital mechanism through which an internal network of knowledge flows develops and is sustained. Jones (2003) details the development of a global corporate culture in international management consulting firms and international banks. Induction programmes provide staff with a means of plugging in to the firm's global network as they immediately acquire a global network of colleagues consisting of the programme's cohort and their wider contacts.

Through the emersion in a global corporate culture consultants have access to a common corporate language that allows them to communicate with the firm's consultants in multiple locations. Consultants are able to translate and adapt knowledge gained from a corporate global network to a specific local assignment. Consultants must retain a local network as well as developing their global network within the firm and beyond. The development of a global culture helps to open national staff to opportunities overseas and promotes a willingness to work across cultures.

Global Training Facilities

In addition to the initial induction programme, consulting staff also engage in regular training programmes, which allow them not only to update their skills but also to update and sustain their social relations with colleagues in the global network. For many of the large consultancy firms, personnel are drawn from across the global network to participate in courses in the firms' global training centres (Jones 2003; Roberts 1998). Importantly, training in a global context helps to create experience based trust between members of the firm. Consultants are able to development reputations within the network for specific expertise and to again knowledge of other experts in the network. That is, they acquire what Lundvall and Johnson (1994) refer to as know-who, which provides them will knowledge of who knows what. Training facilities may also be home to 'centres of excellence'.

Centres of Excellence

Location specific teams of highly mobile personnel enable the knowledge assets embedded in these locations to be shared and exploited internationally. These pools of skills are continuously evolving and may be used in a variety of locations through a combination of computer-mediated communication and movement of key personnel. Centres of excellence are evident in some highly globalized professional business firms. For example, Moore and Birkinshaw (1998) identified a multimedia centre of excellence at Andersen Consulting, which involved a small team of people who worked on multimedia projects and who knew the current technology as well as the firm's experience in the field. They also found a business-to-business marketing centre of excellence that is used to help Mckinsey offices develop marketing efforts with clients in many countries. Similarly, Gentle and Howells (1994) highlight the development of centres of excellence in the European computer services sector. By concentrating expert staff in specific locations, individual and team based knowledge can be used efficiently throughout the global business service firm (Roberts 2002). Locationally fixed staff distributed throughout the company's international network can be supplemented by the temporary mobilization of experts from the company's centres of excellence. In such a way the scarce resources of highly skilled staff are used efficiently. Local knowledge is used to tailor globally developed solutions to the needs of clients in multiple locations.

Global Data Base Systems

Databases may be in the form of standard manuals that are distributed throughout the global organization with a view to harmonizing the service quality and methodological approach. Such mechanisms have been in use for many years; however, these days ICT facilitates the greater use of databases and the access of information resources on a real-time basis. Firms in all sectors of the economy are making use of computer-based databases to improve their use of and access to knowledge and information. The management of knowledge in this way is particularly valuable for knowledge intensive business service firms.

A number of studies of knowledge management in consulting firms highlight the role, and indeed the shortcomings, of firm wide databases (Werr and Stjernberg 2003; Martiny 1998; Dunford 2000). For instance, Dunford (2000, 295-6) details the knowledge management process developed by a number of top firms including McKinsey and Booz, Allen and Hamilton. McKinsey has developed a computer-based Practice Development Network, which contains documents that represent core knowledge in practice areas, a Knowledge Resource Directory, which lists all firm expertise and a Firm Practice Information System, which is a computerized database of all client engagements. Similarly, Booz, Allen and Hamilton have a Knowledge On-Line system, an intranet-based system, which includes a knowledge repository, an expert skills inventory and databases that cover financial reporting, training, recruiting, marketing, human resources, plus numerous external databases.

Such databases can be used across borders (through remote access), whether by consultants on an overseas assignment or located in an overseas office, or indeed, accessed directly by clients. Such technology-based tools are an enabler of knowledge transfer, but they are most effective when combined with direct contact between consultants and their colleagues and clients (Werr and Stjernberg 2003; Martiny 1998).

Client Relations

The production of a customized service involves the transfer of tacit knowledge between the client and producer. Close contact is required to enable the process of learning necessary for the successful transfer of tacit knowledge (Roberts 2000, Wood 2002). A process of co-production involving both the producer and client is facilitated by co-location. It is through this close relationship that sensitive information and knowledge may be shared and a process of knowledge creation can be nurtured (Dawson 2000). Confidentiality, integrity, trust and mutual understanding are essential qualities that characterize a successful client producer relationship. Consequently, client relations are an important mechanism through which knowledge, especially tacit knowledge, may be transferred. Given the importance of the client-producer relationship in the provision of services it is no surprise that client relations are a key source of competitive advantage for business service firms (Aharoni 2000). Key individuals become important interfaces between the consulting firm and the client firm (Jones 2003). Recognizing the significance of these individuals in the coordination of international relations with client firms many business service firms have developed the position of global account managers (Roberts 1998).

Boundary Processes Facilitating Knowledge Transfer

In a sense, globally mobile consultants become knowledge brokers and they distribute knowledge throughout the firm in conjunction with boundary objects (Star and Griesemer 1989). Wenger (1998; 2000) identifies a number of boundary processes through which knowledge can be transferred between communities of practice including brokering, boundary objects, boundary interactions and cross-disciplinary projects. In Table 6.6 these boundary processes are considered in relation to the cross-border delivery of management consulting services both between the consulting firm and overseas client firms and between overseas consulting offices within the consulting firm's global network.

From an examination of the boundary processes relevant to the cross-border supply of consultancy services it can be seen the Mode 3, the supply of a service through the commercial presence of the foreign supplier in the territory of another country, does not in itself constitute a means through which knowledge may be transferred across borders. To facilitate knowledge transfer Mode 3 must be accompanied by one of the other three modes of cross-border service delivery. Intra-firm knowledge transactions are then essential for the international provision of

Table 6.6 Boundary processes in management consultancy cross-border service delivery modes

Boundary Processes	Examples from the Cross-Border Delivery Management Consultancy Services	Mode of Cross-Border Services Delivery
Boundary Objects	Traded goods in the form of books, reports, videos, firm specific methodologies and management principles (e.g. Total Quality Management, Business Process Reengineering) embodied in manuals, etc. Access to cross border databases.	Mode 1 Cross-border trade
Boundary Interactions	Participation in cross-border training programmes, socialization activities, both intra and extra firm. Promotional events, e.g. engaging in local and national activities and events in the wider national and international community.	Modes 2 & 4 Temporary movement of customers and/or producers
Cross-disciplinary Projects	Projects involving management consultants and their client firms' staff, research collaboration between managers and client firms across national borders. Cross-border intra-firm teams working on a common client project.	Modes 2 & 4 Temporary movement of customers and/or producers
Brokering	Movement of consultants on temporary assignments and short-term secondments. Global account managers facilitating cross-border client relationships. Personnel delivering training programmes across borders or in global training facilities. Key experts working across-borders Key individuals acting as knowledge brokers at the interface between the firm and its clients.	Mode 4 Temporary movement of producers

Table 6.7 Modes of international service delivery in consulting services

Mode of service delivery	Examples	Significance for internationalisation of consulting services	Nature of knowledge transfer	Dominant direction of knowledge flow
Cross-border trade	Client access to on-line database services, transfer of files and documents including reports and manuals	Cross-border trade most common in intra-firm trade with intermediate products	Codified knowledge	From home country to overseas clients or affiliate
Movement of customers	Workshops for clients, e.g. training on the premises of consulting firm in the home country	Of minor importance	Tacit and codified knowledge	From home country to overseas clients
Temporary movement of producers	Consultants working on location, e.g. monitoring the implementation of projects	Predominant in engineering and management consultancy .	Tacit and codified knowledge	From home country to overseas clients
Delivery through the establishment of affiliates abroad	Local affiliates acting as consultants, occasionally with direct support from parent firm	Importance growing because of mergers	Tacit and codified knowledge	From home country to overseas clients and from overseas to home country: Two way flow of knowledge However, where the affiliate is completely autonomous no knowledge flow need occur
Intra-firm service transactions	Training, intermediate inputs including access to data bases and centres of excellence, staff rotation.	Large consulting firms have sophisticated knowledge management systems	Tacit and codified knowledge	Interactive flows of intra-firm knowledge across national boundaries between home country and affiliates and between overseas affiliates: Multi-directional flow of knowledge

consultancy services, even when this occurs through an overseas affiliate. When an affiliate provides a service without intra-firm knowledge transactions then there is no cross-border knowledge flow. Knowledge used in the delivery of such services is indigenous to the affiliate's location. Although the establishment of an overseas affiliate will require an initial knowledge flow, perhaps through the location of expatriate staff in the overseas location, the subsequent delivery of services does not require a cross-border flow of knowledge unless accompanied by at least one of the other modes of service delivery.

A Framework for the Cross-Border Transfer of Knowledge Through Management Consultancy Services

From the analysis presented in the previous section it is possible to construct a framework that illustrates the various mechanisms through which knowledge transfer occurs in the cross-border delivery of management consultancy services. Importantly, these interactions are dynamic and interdependent. This framework is captured in Table 6.7. The modes of cross border management consultancy service delivery are detailed together with the nature of knowledge associated with the mode and the dominant directional flow of the knowledge.

A comprehensive conceptualization of the cross-border knowledge transfer facilitated through the international delivery of management consultancy services requires the recognition of intra-firm trade in knowledge as an essential component of cross border transactions. Intra-firm service transactions must be added to the four modes of cross border supply identified by GATS (Table 6.4). An important characteristic of intra-firm trade is that it can facilitate the multi-directional flow of knowledge throughout the global network of a consultancy firm. While trade and the accompanying knowledge flows associated with the international delivery of knowledge intensive services is traditionally viewed as one-way, the delivery of services through the establishment of an affiliate may give rise to a two-way flow of knowledge, whether intra-firm or inter-firm (i.e. directly with a client in a number of locations). However, it is also the case that when an affiliate serves a client in its home market, it may utilize independently created knowledge rather than draw on parent firm's knowledge base. For instance, the delivery of a service may depend on knowledge of location specific regulation or business conditions. Consequently, an affiliate may draw on the firm's knowledge base in order to supply certain services, but at times it may draw solely on its own knowledge base.

Furthermore, recognition of intra-firm trade reveals the full complexity of knowledge flows within global knowledge intensive service firms. One or two way knowledge flows may be accompanied by multi-directional flows of knowledge. Given that consultancy firms deliver a range of both customized and standardized services it is reasonable to assume that those services that are standardized may include a greater proportion of knowledge drawn from the firm's global network. On the other hand, services that are customized will depend more on locally produced knowledge. Importantly, though, the provision of a service will often include a combination of firm-specific knowledge and location specific knowledge together with client knowledge.

In terms of understanding the internationalization of management consultancy services, the analysis of the cross-border delivery of knowledge intensive services suggests that a more detailed appreciation of the internationalization of these services would benefit from the consideration of international transactions alongside the firm's global knowledge management strategy.

Conclusion: An Agenda for Future Research

This chapter has examined the internationalization of management consultancy services, as an example of KIBS, with the aim of identifying the specific issues for internationalization that face this sub-sector, and particularly, conceptual issues concerning the cross-border delivery of knowledge-intensive services. While it is difficult to provide a precise definition of management consultancy services because of the dynamic nature of the sector, it is, nevertheless, evident that the sector is growing rapidly. The sector is becoming increasingly internationalized. Management consulting involves the creation and dissemination of knowledge, providing an example of a knowledge intensive service. After having examined the cross-border transfer of knowledge in a general sense, the role of management consultancies in the cross-border transfer of knowledge was considered. From this a framework was proposed which highlights the importance of intra-firm transactions in the facilitation of the cross-border supply of knowledge intensive services.

This analysis of the internationalization of management consultancy services provides a number of important insights into the internationalization of knowledge intensive services. First, the analysis highlights the significance of intra-firm transactions, including trade, which are essential for the cross-border supply of knowledge intensive services. The establishment of an overseas presence, whether through, for instance, mergers, acquisitions or joint ventures, while facilitating the cross-border delivery of knowledge intensive services is not sufficient to enable the opening up of a continuous cross-border flow of knowledge. An overseas presence must be accompanied by other modes of service delivery, in the form of intra or extra firm transactions, if a continuous flow of knowledge is to be sustained. The existence and importance of intra-firm trade in services has already been identified (Roberts 1998 1999; Coffey and Polese 1987). However, what emerges from the analysis here is the need for intra-firm trade and other transactions to be considered in terms of the nature of the knowledge transaction. There are then important links to be made between a firm's knowledge management strategy and its internationalization strategy. Moreover, it is clear that when exploring knowledge intensive services international transactions in the form of trade, investment and the movement of people cannot be thoroughly examined without taking into account the organizational development of the firm. Nevertheless, it is important to note that an affiliate may produce knowledge independently in the local market requiring no cross-border knowledge flow to service clients in that market.

Furthermore, a full understanding of cross-border knowledge intensive service delivery requires an appreciation of the relationship between the producer and the client. For the client is also an intricate element in the knowledge creating and dissemination dynamic associated with a knowledge intensive service firm. Consequently, clients firms should be incorporated, in some way, into the consultancy firm's cross-border knowledge management strategy.

The key assets of management consulting firms include intellectual capital, reputational capital and relational capital, all of which are embodied in individuals. As a result, many new consultancy firms arise when consultants leave a firm to establish their own business making use of intellectual, reputational and relational capital gained with their former employer. This has important implications for the development, management and retention of consultancy staff. As note above, certain consultants become key players developing close relations with client firms and acting as knowledge brokers in an international context by becoming an important element of the interface between the firm and its global clients.

In summary four key areas emerge as important directions for future research. Firstly, a greater understanding of intra-firm knowledge transactions, including their extent and significance is required. Secondly, the development of international knowledge management strategies and their impact on the international organizational structure of knowledge intensive service firms deserves attention. Thirdly, a deeper appreciation of client relations in an international context and their role in knowledge transfer is required. Finally, the role of key personnel as brokers of cross border knowledge flows merits investigation. Clearly further research is required to develop our understanding of these issues. This initial investigation into the internationalization of management consultancy services has provided some important insights into the cross-border delivery of knowledge intensive services. Further theoretical and empirical research is required to unravel the full complexities of the internationalization of knowledge intensive business services including management consultancy services.

References

Abrahamson, Eric (1996), 'Management Fashion', *Academy of Management Review* 21: 1, 254–85.

Aharoni, Yair (2000), 'The Role of Reputation in Global Professional Business Services', in Aharoni, Yair and Nachum, Lilach (eds), *Globalization of Services: Some Implications for Theory and Practice* (London: Routledge), 125–141.

Aharoni, Yair (1999), 'Internationalization of Professional Services: Implications for Accounting Firms', in Brock, D., Powell, M. and Hinings, C.R. (eds), *Restructuring The Professional Organization: Accounting, Healthcare and Law* (London and New York: Routledge), 20–40.

Aharoni, Yair (ed.) (1993), *Coalitions and Competition: The Globalization of Professional Business Services* (London: Routledge).

Aharoni, Yair and Nachum, Lilach (eds) (2000), *Globalization of Services: Some Implications for Theory and Practice* (London: Routledge).

Andersen, Birgitte, Howells, Jeremy, Hull, Richard, Miles, Ian and Roberts, Joanne (2000), *Knowledge and Innovation in the New Service Economy* (Cheltenham: Edward Elgar).

Antonelli, Cristiano (1999), *The Microdynamics of Technological Change* (London and New York: Routledge).

Beaverstock, Jonathan, V. (2004), '"Managing Across Borders": Knowledge Management and Expatriation in Professional Service Legal Firms', *Journal of Economic Geography* 4: 2, 157–179.

Beaverstock, Jonathan V., Smith, R.G. and Taylor, P.J. (1999), 'The Long Arm of the Law: London Law Firms in a Globalising World', *Environment and Planning A* 31, 1857–1876.

Bryson, John (2002), '"Trading" Business Knowledge Between Countries: Consultants and the Diffusion of Management Knowledge', in Cuadrado-Roura, J.R., Rubalcaba-Bermejo, L. and Bryson, J.R. (2002), *Trading Services in the Global Economy* (Cheltenham: Edward Elgar), 175–190.

Bryson, John R., Daniels, Peter W. and Warf, Barney (2004), *Service Worlds: People, Organizations, Technologies* (London and New York: Routledge).

Clark, Timothy and Fincham, Robin (2002), 'Introduction: The Emergence of Critical Perspectives on Consulting', in Clark, Timothy and Fincham, Robin (eds), (2002), *Critical Consulting: New Perspectives on the Management Advice Industry* (Oxford: Blackwell), 1–20.

Coffey, W.J. and Polese, M. (1987), 'Intrafirm Trade In Business Services: Implications For The Location Of Office-Based Activities', *Papers Of The Regional Science Association* 62, 71–80.

Commission of the European Communities (1998), *The Contribution of Business Services to Industrial Performance: A Common Policy Framework*, Brussels, 21.9.1998, COM (1998) 534 final.

Commission of the European Communities (2000), *The Panorama of European Business*, Luxembourg: Office for Official Publications of the European Communities.

Commission of the European Communities (2001), *Barriers to Trade in Business Services*, Centre for Strategy & Evaluation Services.

Cuadrado-Roura, J.R., Rubalcaba-Bermejo, L. and Bryson, J.R. (2002), *Trading Services in the Global Economy* (Cheltenham: Edward Elgar).

Czerniawska, Fiona (2004), *The UK Counsulting Industry 2003/4*, London: Management Consulting Association in conjunction with PPP and IRN Research.

Daniels, P.W. (1993), *Service Industries in the World Economy* (Oxford: Blackwell).

Daniels, P.W., Thrift, N.J. and Leyshon, A. (1989), 'Internationalization of

Professional Producer Services: Accountancy Conglomerates', in Endwick, Peter, *Multinational Service Firms* (London: Routledge), 79–106.

Dawson, R. (2000), *Developing Knowledge-Based Client Relationships: The Future of Professional Services* (Oxford: Butterworth-Heinemann).

Dunford, R. (2000), 'Key Challenges in the Search for the Effective Management of Knowledge in Management Consulting Firms', *Journal of Knowledge Management* 4: 4, 295–302.

FEACO (2002), *Survey of the European Management Consultancy Market*, FEACO: Brussels, available at: http://www.feaco.org (last accessed 15th June 2004).

Ferguson, M. (2001), *From Efficiency Engineer to Consultant, From Productivity to Strategy: The History of Management Consulting in Britain* (Aldershot: Ashgate).

Gentle, C. and Howells, J. (1994), 'The Computer Services Industry: Restructuring for a Single Market', *Tijdschrift voor Economische en Sociale Geografie*, 85: 4, 311–321.

Glückler, J. and Armbrüster, T. (2003), 'Bridging Uncertainty in Management Consulting: The Mechanisms of Trust and Networked Reputation', *Organization Studies*, 24: 2, 269–297.

Greiner, L. and Metzger, R. (1983), *Consulting to Management* (Englewood Cliffs, NJ: Prentice-Hall).

Grosse, R. (1996), 'International Technology Transfer in Services', *Journal of International Business* 27: 4, 781–800.

Howells, J. and Roberts, J. (2000), 'Global Knowledge Systems in a Service Economy', in Andersen, Birgitte, Howells, Jeremy, Hull, Richard, Miles, Ian and Roberts, Joanne, *Knowledge and Innovation in the New Service Economy* (Cheltenham: Edward Elgar), 248–266.

Jones, A. (2003), *Management Consultancy and Banking in an Era of Globalization* (Basingstoke: Palgrave Macmillan).

Kennedy Information (2004), *The Global Consulting Marketplace 2004–2006: Key Data, Trends & Forecasts*, Kennedy Information Inc. available at: http://www. consultingcentral.com/reports/GlobalConsultExecSum.pdf (last accessed 16th June 2004).

Kieser, A. (1997), 'Rhetoric and Myth in Management Fashion', *Organization* 4: 1, 49–74.

Kipping, M. (1996), 'The US Influence on the Evolution of Management Consultancies in Britain, France, and Germany Since 1945', *Business and Economic History* 24: 1, 112–123.

Kipping, M. (2002), 'Trapped in Their Wave: The Evolution of Management Consultancies', in Clark, T. and Fincham, R. (eds), *Critical Consulting: New Perspectives on the Management Advice Industry* (Oxford: Blackwell), 28–49.

Kipping, M. and Sauviat, C. (1996), *Global Management Consultancies: Their Evolution and Structure*, Discussion Papers in International Investment and business Studies, Series B Vol. IX, Department of Economics, University of Reading.

Kozul-Wright, Z. and Howells, J. (2002), *Changing Dynamics of Global Computer Software and Services Industry: Implications for Developing Countries* (Geneva: United Nations).

Leslie, D.A. (1995), 'Global Scan: The Globalization of Advertising Agencies, Concepts, and Campaigns', *Economic Geography* 71: 4, 402–426.

Lundvall, B-Å. and Johnson, B. (1994) 'The Learning Economy', *Journal of Industry Studies* 1: 2, 23–42.

Management Consultancy, (2004), 'A Booming Market, But the Outlook Is Less Certain...', May, pp. 5–7.
(Available at: http://www.managementconsultancy.co.uk/ accessed 16th June 2004).

Martiny, M. (1998), 'Knowledge Management at HP Consulting', *Organizational Dynamics* 27: 2, 71–77.

Mattelart, A. (1992), *Internationalization of Advertising* (London: Routledge).

McKenna, C.D. (1995), 'The Origins of Modern Management Consulting', *Business and Economic History*, 24: 1, 51–58.

Miles, I., Kastrinos, N., Flanagan, K., Bilderbeek, R., Hertog, P., Huntink, W. and Bouman, M. (1995), *Knowledge-Intensive Business Services: Users, Carriers and Sources of Innovation*, EIMS Publication No. 15, Innovation Programme, Directorate General for Telecommunications, Information Market and Exploitation of Research, Commission of the European Communities, Luxembourg.

Miozzo, M. and Miles, I. (eds) (2002), *Internationalization, Technology and Services*, (Cheltenham: Edward Elgar).

Moore, K. and Birkinshaw, J. (1998), 'Managing Knowledge in Global Service Firms: Center of Excellence', *Academy of Management Executive* 12: 4, 81–92.

Nachum, L. (1999), *The Origins of the International Competitiveness of Firms: The Impact of Location and Ownership in Professional Service Industries* (Aldershot: Edward Elgar).

Nicolaides, P. (1989), *Liberalizing Service Trade* (London: Routledge).

Noyelle, T.J. and Dutka, A.B. (1988), *International Trade in Services: Accounting, Advertising, Law, and Management Consulting* (Cambridge, Massachusetts: Ballinger).

Nusbaumer, J. (1987), *Services in the Global Market* (Boston: Kluwer Academic Publishers).

OECD (1989), *The Internationalization of Software and Computer Services*, Information Computer Communications Policy Series No. 17, OECD: Paris.

OECD (1990), *Trade in Information, Computer and Communication Services*, Information Computer Communications Policy Series No. 21, OECD: Paris.

OECD (1999), *Strategic Business Services*, Paris: OECD.

OECD (1997), *Services Statistics on Value Added and Employment*, Paris: OECD.

Office of National Statistics (2003), *UK Balance of Payments: The Pink Book*, London: Office of National Statistics.

Office of National Statistics (2005), *UK Balance of Payments: The Pink Book*, London: Office of National Statistics.

O'Farrell, P.N., Moffat, L. and Wood, P.A. (1995), 'Internationalization by business services: a methodological critique of foreign-market entry-mode choice', *Environment and Planning A* 27, 683–97.

Perry, M. (1990), 'The internationalization of advertising', *Geoforum* 21, 35–50.

Polanyi, M. (1958), *Personal Knowledge: Towards a Post-Critical Philosophy*, (London: Routledge & Kegan Paul).

Riddle, D. (1986), *Service-Led Growth: The Role of the Service Sector in World Development* (New York: Praeger).

Roberts, J. (1998), *Multinational Business Service Firms: The Development of Multinational Organizational Structures in the UK Business Services Sector* (Aldershot: Ashgate).

Roberts, J. (1999), 'The Internationalization of Business Service Firms: A Stages Approach', *The Service Industries Journal* 19: 4, 68–88.

Roberts, J. (2000), 'From Know-how to Show-how? Questioning the Role of Information and Communication Technologies in Knowledge Transfer?, *Technology Analysis and Strategic Management*, 12: 4, 429–443.

Roberts, J. (2002), 'From Market to Resource-Oriented Overseas Expansion: Re-examining a Study of the Internationalization of UK Business Service Firms', in Miozzo, Marcela and Miles, Ian (eds), *Internationalization, Technology and Services* (Cheltenham: Edward Elgar), 161–183.

Roberts, J. (2003), 'Competition in the Business Services Sector: Implications for the Competitiveness of the European Economy', *Competition & Change* 7: 2/3, 127–146.

Sampson, G.P. and Snape, R.H. (1985), 'Identifying the Issues in Trade in Services', *World Economy* 8: June, 171-181.

Shelp, R. K. (1981), *Beyond Industrialization: Ascendancy of the Global Service Economy* (New York: Praeger).

Spar, D. (1997), 'Lawyers Abroad: The Internationalization of Legal Practices', *California Management Review*, 39, 8–28.

Star, S.L. and Griesemer, J.R. (1989), 'Institutional Ecology, "Translations" and Boundary Objects: Amateurs and Professionals in Berkeley's Museum of Vertebrate Zoology 1907–39', *Social Studies of Science* 19: 3, 387–420.

Stern, R.M. and Hoekman, B.M. (1988), 'Conceptual Issues Relating to Services in the International Economy', in C.H. Lee and S. Naya (eds), *Trade and Investment in Services: in the Asia-Pacific Region*, Pacific and World Studies No.1, (Boulder, Colo.: Westview Press).

UNCTAD (1983), *Production and Trade in Services, Policies and Their Underlying Factors Bearing Upon International Service Transactions*, (TD/B/941) (New York: United Nations).

UNCTAD (2002), *The Tradability of Consulting Services: And its implications for developing countries* (New York: United Nations).

Ward, K. (2001), *Global Restructuring in the Temporary Help Industry* (New York: Rockefeller Foundation).

Ward, K. (2004), 'Going Global? Internationalization and Diversification in the

Temporary Staffing Industry', *Journal of Economic Geography* 4: 3, 251–273.

Wenger, E. (1998), *Communities of Practice: Learning, Meaning, and Identity* (Cambridge: Cambridge University Press).

Wenger, E. (2000), 'Communities of Practice and Social Learning Systems, *Organization* 7: 2, 225–246.

Werr, A. and Stjernberg, T. (2003), 'Exploring Management Consulting Firms as Knowledge Systems', *Organization Studies* 24: 6, 881–908.

Wood, P. (ed.) (2002), *Consultancy and Innovation: The business Service Revolution in Europe* (London: Routledge).

WTO (1999), 'An Introduction to the GATS' WTO Secretariat, Trade in Services Division, October 1999: www.wto.org/english/tratop_e/serv_e/gsintr_e.doc (12/7/02).

Chapter 7

The Internationalization of Europe's Contemporary Transnational Executive Search Industry

Jonathan V. Beaverstock, Sarah J.E. Hall and James R. Faulconbridge

Conceptual and empirical lacunae remain in understanding the growth of labor-related knowledge-intensive business services (KIBS) in Europe. Such growth has been partially addressed by Peck, Theodore and Ward (see for example, Ward 2004; Peck and Theodore 2001) in their analysis of the internationalization of temporary staffing agencies. In professional and managerial executive recruitment, however, little geographical work has explored the pivotal roles of transnational executive search firms ('headhunters'), since Boyle et al. (1996) and Cuthbertson (1996) analysed the European industry in the early 1990s. Headhunters are intermediary KIBS specializing in finding and recruiting the most suitable individuals for senior managerial or board level vacancies. This normally means targeting individuals who are not actually in the market for a new job (see for example, Britton et al. 2000; Finlay and Coverdill 1999; Jones 1989). The industry has a relatively short institutional history dating back to the post-war economic boom of the 1940s in the USA. It was between 1950 and 1970 that executive search firms proliferated and the 'big four' were established (Heidrick and Struggles [1953], Spencer Stuart [1956], Russell Reynolds [1969] and Korn/Ferry [1969]), which quickly became leaders in the USA and later in Europe (Britton et al.,1995; Jenn 2005). In Europe, at this time, indigenous firms like Egon Zehnder (1964 – Zurich), Alexander Hughes (1965 – London), Goddard Kay Rogers (1970 – London), Whitehead Mann (1976 – London), Norman Broadbent (1983 – London) and Saxton Bampfylde (1986 – London) all entered the market to compete with the US firms who had set the benchmark from the 1970s, and thus the European industry was developed, first through London during the halcyon days of the 1980s (Bryne 1986; Cuthbertson 1996; Jenn 1993; Jones 1989).

By 2004 Europe had become the most important global market for headhunting. According to Jenn (2005), for the top 20 global headhunting firms Europe accounted for 49 per cent of total worldwide revenue, compared to 37 per cent for the USA and 11 per cent for Pacific-Asia. In terms of office distribution Europe accounted for 43 per cent of total worldwide offices (343), as compared to 25 per cent (196) in the USA. In this context, the aim of this chapter is to explore the recent internationalization and regionalization of Europe's transnational headhunting industry. By drawing upon the work of selected headhunting commentators from business and consultancy (Britton et al. 2000; Finlay and Coverdill 2002; Jenn 2005) and detailed analyses of headhunting transnational firm-specific data sourced from firm world-wide web sites, we detail the contemporary strategies that headhunters use to consolidate in Europe and the USA and penetrate new markets, especially in Latin America and the Asia-Pacific. The rest of this chapter is organized into six parts. Following this introduction, we discuss the emergence of the headhunting as a *bona fide* industry in the 1950s and discuss its practices, to understand why internationalization has been mainly through foreign direct investment (FDI). In part three, we briefly revisit the resource-based theoretical writings of commentators like John Dunning (1993; Dunning and Norman 1983, 1987) and Nachum (1999), which inform us as to why KIBS engage in international production through FDI and a direct presence in an international market. This is followed in part four by a focused discussion of the internationalization of the leading European transnational headhunters in the world. Part five maps and traces the geographies of the worldwide office networks of Europe's leading sixteen transnational firms from 1992 to 2005, which is followed in part six by three vignettes of different organized firms in internationalization: wholly-owned (Egon Zehnder); networked (The Globe Search Group); and, hybrid (Amrop Hever Group). Finally, the chapter concludes by discussing how our new research on Europe's leading headhunting transnational corporations can deepen the understanding of internationalization and regionalization strategies of KIBS in contemporary globalization.

The Emergence of Headhunting as a European and Global Industry

The headhunting industry emerged in the USA during the post-war economic boom of the 1940 and 1950's. Demand for executives to steer expanding and newly emerging organizations grew rapidly and, for the first time, firms were unable to fill management positions through internal labor markets (Jones 1989). Demand for externally recruited talent increased at rates exceeding the readily available supply of executives seeking new positions. Management consultants and accountants in particular recognized the opportunity this phenomenon offered for a new type of professional service, the executive search function. Headhunting, as a complementary service function, began to emerge within existing management consultant and accounting firms (for example, Arthur Andersen). This rapidly evolved as the market for executive search grew. Individuals and teams began to leave their original employers to develop stand alone executive search organizations dedicated solely to serving this newly emerging market. Between 1950 and 1970 leading US executive search firms expanded domestically (for example opening offices in New York, Chicago, Los Angeles) and the market soon became dominated by the activities of the 'big four' (Heidrick and Struggles [1953], Spencer Stuart [1956], Russell Reynolds [1969] and Korn/Ferry [1969]).

Within Europe, the 'Europeanization' of the industry of headhunting (see Britton et al. 1995) emerged principally through the opening of offices in London by the 'big four' (Heidrick and Struggles in 1968, Spencer Stuart in 1961, Russell Reynolds in 1971 and Korn/Ferry in 1973), and the mimicking of their functions and practices by indigenous firms in London and beyond (for example, Paris, Brussels, Zurich, Frankfurt). London and Europe's burgeoning headhunting industry, composed of US firms with a large market share and local firms driven by American practice and management styles, contributed significantly to a sea change in the mediation of executive recruitment in the labor market. First, they contributed to the decline in the belief that all employees should 'serve their time' and be trained to act as a suitable manager for the organization. Acknowledgement grew of the value of 'mobile talent': individuals bought in from outside because of their skills and expertise (see Sennett 1998 on the demise of professional jobs for life). Second, they began to circumvent the use of an 'old-boy network' to recruit senior executives, something peculiar to London, which was inefficient and unsuitable in the 'new' business environment. Instead it was recognized that a more scientific and objective strategy was needed. As Jones (1989, 19) notes, 'business has become too complicated, and the stakes have become too high, for a board chairman who needs executive talent to rely on his [sic] friends or his friends' recommendations. He wants the best man, not the best man visible'. Combined, these changes created the opportunity for a new mode of executive recruitment in Europe: a mode mediated through executive search firms specialising in seeking out and matching talent to vacancies.

From the 1980s onwards this growth in Europe's executive search industry was reflected in companies' revenue and profit figures. For example, Boyden International saw billings in the UK rise from under £300,000 in 1982 to £1.06m in 1987 (Jones, 1989, 24). This rise of over 250 per cent in five years was spectacular and was reflected by similar levels of growth in other European offices. The emergence of the rhetoric of a 'knowledge economy' from the 1990s onwards (see Department for Trade and Industry, 1998; Leadbeater, 1999), only served to reinforce this growth. As *Fast Company* (2000, 44) noted about the predominant logic in this period, 'it's hard to argue with the idea that the company with the best talent wins'.

The Firms and the Service

The newly internationalized executive search industry in Europe has been dominated by sixteen transnational firms (Table 7.1). These are 'retained' organizations: organizations selected by clients who have an executive position to fill and who pay a fee (normal practice is for the fee to be 33 per cent of the recruited candidate's salary) to manage the search and selection of a suitably talented individual (Jenn, 1993). This differs from 'contingent headhunters' who have to 'fight' to win a client's business based on the quality of the shortlist provided (i.e. they have to do the search to win the business whereas retained headhunters only search after being selected). The latter are normally involved in search and recruitment for non-executive vacancies, typically administrative or customer service staff, whereas

Table 7.1 Europe's top sixteen transnational headhunting firms (ranked by 2004 European revenue), 2005

Firm (founded)	Revenue ($m) Europe	Revenue ($m) Worldwide	Global HQ	Structure	Offices Europe	Offices Worldwide	World Ranking
1. Korn/Ferry International (1969)	132	402	Los Angeles	Owned	23	73	1
2. Heidrick & Struggles Int'l (1953)	128	375	Chicago	Owned	22	59	2
3. Ray & Berndtson (1965)	115	147	New York	Hybrid	28	48	6
4. Russell Reynolds Ass. (1969)	110	268	New York	Owned	12	33	5
5. Spencer Stuart (1956)	109	362	Chicago	Owned	19	49	3
6. Amrop-Hever (2000)	88	135	Brussels	Hybrid	40	78	7
7. Whitehead Mann Group (1971)	52	61	London	Owned	7	10	11
8. Globe (1997)	44	76	London	Network	10	15	8
8. IIC Partners (1986)	44	75	Alberta	Hybrid	26	53	9
10. TransSearch (1982)	41	70	Paris	Hybrid	31	67	10
11. Signium Int. (1972)	34	43	Crystal Lake	Hybrid	18	30	14
12. Eric Salmon & Partners (1990)	14	16	Paris	Owned	4	5	17
13. A.T.Kearney (1946)	13	45	Chicago	Owned	12	27	13
14. Stanton Chase Int. (1990)	12	39	Dallas	Hybrid	17	59	15
15. Egon Zehnder Int'l (1964)	NA	336	Zurich	Owned	28	59	4
15. Boyden (1946)	NA	NA	New York	Owned	28	66	NA

NA Data not available.
Source: Firm worldwide web sites, accessed 07/12/05; adapted from Jenn (2005).

retained firms exclusively deal with elite executive search (Jenn, 1993; 2005). In particular, it is important to distinguish between the two dominant forms of retained international headhunting firm. On the one hand there are specialist, 'born-global' boutiques that concentrate on headhunting in one sector, such as financial services. For example, Boyden International Ltd based in London, founded in 1966, employed just eight staff in 1993. The firm concentrates on the retail and consumer goods market only (Jenn 1993). On the other hand, there are fully integrated transnational corporations in the most traditional sense of the word, a category that includes many of the global industry leaders such as Korn/Ferry International and Spencer Stuart. For example, Korn/Ferry has over 400 partners in addition to its retained consultants and researchers (Jenn 2005). The retained firms fundamentally offer the same type of service, search and selection, whilst each having their own unique cultures and styles (Jones, 1989) and organizational forms.

The Standard Search and Selection Sequence.

A review of the extant business literatures and specific firms' websites suggest that the search and selection *modus operandi* of headhunting has five main stages (Finlay and Coverdill 1999; 2000; Gurney 2000; Jenn 2005; Jones 1989; www.kornferry. com):

- First, the 'beauty parade' or 'shootout' where several firms are asked to pitch for the client's work. Here, rather than producing a list of candidates as contingent firms do, retained firms sell themselves based on their competencies, knowledgeable employees and past experience. The aim is to convince the client they will find the most suitable candidate.
- Second, after a firm is selected, the task definition stage follows. Here the headhunter works with the client to identify: the key characteristics and culture of the organization and how candidates must fit within these; the ideal characteristics, competencies and strengths of a candidate; a search strategy that has the required direction and focus in order to find the best candidates.
- Third, short-listing takes place based on potential candidates: (a) in the search firm's candidate database; and (b) known to the consultants involved in the project. The division of labor in this process is split between: researchers, whose sole role is to produce a list of candidates, investigate their credentials and gain insights from past or present colleagues of the individual; and consultants (partners in the executive search firm) who manage the relationship with the client and assess and narrow down the shortlists researchers produce. Once consultants have identified the top targets these candidates are then discretely contacted and interrogated about their interest in such a vacancy (whilst maintaining the anonymity of the client), normally by researchers.
- Fourth, the client reviews the shortlist and interview meetings are established between the client and a selected number of candidates. Here the role of the headhunter is to ensure these meetings are kept confidential and no two

candidates are aware of one another's involvement in the search/recruitment process. If the client is unhappy with the candidates after interview, phase three begins again.

- Finally, 'integration handholding' occurs as salary negotiations and the integration of the successful candidate into the client's firm is smoothed by support given from the executive search firm.

Combined, these stages are designed to ensure the client's need for skilled elite labor is met in the most suitable way. An unsuccessful assignment would be one where the successful candidate leaves the client's firm within the first year, for whatever reason. Overall, the role of the headhunter is to ensure the candidate: (a) has the skills needed to fulfil the role; and (b) will fit in and feel comfortable in the position they are selected for in the client firm.

Service Delivery Through Experts with Functional and Practice Specialism

Executive search consultants are a new breed of professional who are both well educated (often to MBA level) and expert in 'functional' and 'industry' fields. 'Functional' expertise refers to the type of search (for example, management executive or board member) whilst 'industry' expertise refers to one of five common practice areas (finance and business services; technologies and telecoms; consumer goods/service; industrial; and life sciences/healthcare) (The Executive Grapevine 2004; Jenn 2005). This means all clients receive a search service that is tailored to identify only the candidates with suitable skills and experience relevant to the client's industry and current vacancy.

All transnational firms in Europe offer search and selection as well as various forms of management consultancy. In an ironic twist, headhunting organizations, which emerged from management consultancy firms, are now offering add-on services such as human resource consultancy and 'leadership' services (Jenn 2005). The firms claim such services complement the recruitment of talented individuals by allowing the most effective management of human (intellectual) capital through consultancy in relation to performance benchmarking and measurement, succession planning and the effective integration of teams after mergers. As the home page of Heidreick and Struggles' website reads, '[A]s innovators we are actively redefining top-level search to encompass complementary services that help build strong companies and the leaders of tomorrow. Our comprehensive approach to leadership acquisition, assessment and development enables us to help our clients build high-performance leadership teams' (http://www.heidrick.com/default.aspx, accessed 20/01/06). However, whilst this additional, value-adding, professional advice is now central to the offering of such firms, the basic headhunting/executive search element remains the key part of their business.

Of course, in essence, the search and recruitment process is one that internal human resource departments could, and traditionally have, fulfilled. However, the emergence of an international headhunting marketplace in Europe is the result of a number of overlapping forces from both demand and supply sides that have made it profitable for firms such as the 'big four' to operate integrated, transnational, executive search networks.

The Ownership-Location-Internalization Paradigm and the Internationalization of KIBS

John Dunning's (1993; Dunning and Norman 1983; 1987) resourced-based Ownership-Location-Internalization (OLI) paradigm has been one major approach used to explain why KIBS firms engage in international production through FDI, thereby becoming TNCs. This approach is widely quoted in economic geography and international business literatures (Daniels et al. 2004; Dicken 2003). According to Dunning and Norman (1983) resource-based and import substituting service firms (for example, reinsurance, executive search, accounting, consultancy, investment banking) will only engage in FDI, principally through wholly-owned subsidiaries or partnerships, where they have OLI specific advantages over local firms. Ownership specific advantages are based on the firm's specialist knowledge, international reputation and accessibility to the market. Location specific advantages are based on the importance of having on the spot contact with clients (face-to-face), skilled labor and easy access to the market, as well as local information. Internalization specific advantages exist, according to John Dunning, in the form of the ability to protect and exploit O and L specific advantages, and to control the quality of the service delivered to the client. If they do not have such advantages over local firms then the penetration of foreign markets is best served through franchises, joint-ventures or international trade in services (Bryson et al., 2004)

The OLI paradigm is a very important approach to explain the relevant influences on the internationalization of different KIBS firms via FDI (Table 7.2). However, it must be noted that the strategies and modes of internationalization are complex within and between different KIBS sectors. For example, Roberts (1999) notes how producer services firms often take a 'stages approach' using different organizational forms at different stages of internationalization. Penetration into international markets is initiated by exports in services through, for example, 'embodied services' (for example, report writing), 'wired services' (for example, telephone conversation) or 'transhuman exports' (for example, personnel travelling to an overseas market to present a report), and then is deepened by FDI and the establishment of an overseas presence (for example, wholly owned subsidiary, joint-venture, franchise), which is further enhanced through intra-firm trade in services (Roberts 1999, 74 and 76–77). In addition, the complexities of internationalization are readily discussed in many of the recent empirically-rich publications which have focused on examples such as: advertising (Nachum 1999); computing services (Coe 1997); investment banking (Beaverstock 2006); and legal (and general professional) services (Morgan and Quack, 2005a, 2005b).

Table 7.2 Ownership, location and internalization advantages in selected transnational KIBS

	Ownership advantages	Location advantages	Internalization advantages	Organizational form
Accounting	Access to transnational clients. Experience of standards loosely required. Brand image of leading accounting firms.	On-the-spot contact with clients. Accounting tends to be culture-sensitive. Adaptation to local reporting standards and procedures. Oligopolistic interaction.	Limited interfirm linkages. Quality control over (international) standards. Government insistence on local participation.	Mostly partnerships or individual proprietorships. Overseas subsidiaries organized, little centralized control. Few joint ventures.
Consulting	Access to market. Reputation, image, experience. Economies of specialization in particular levels of expertise etc., skills, countries.	Close contact with client. The provision is usually highly customer-specific. TNC clients might deal with headquarters. Mobility of personnel.	Quality control. Fear of under-performance by licensee. Knowledge sometimes very confidential and usually idiosyncratic. Personnel coordinating advantages.	Mostly partnerships or 100 per cent subsidiaries. A lot of movement of people.
Legal Services	Access to transnational clients and knowledge of their particular needs. Experience and lawyers reputation.	Need for face-to-face contact with clients. Foreign customers may purchase services in home country. Need to interact with other local services. Restrictions on use of foreign barristers in courts. Extent of local infrastructure.	Many transactions are highly idiosyncratic and customer specific. Quality control. Need for understanding of local customers and legal procedures.	Some overseas partnerships, but often services are provided via movement of home country to clients.

Source: adapted from Dunning (1993, 272).

The Internationalization of Headhunting

As we have noted, the internationalization of executive search firms into Europe began in the 1960s (Britton et al., 1997; Jenn, 1993). However, recently a new form of transnational integration has emerged in both the executive search marketplace and the executive search firms operating within it. Whereas in the initial phases of internationalization (1960–1980) each office of an international firm served clients through search for executive labor *within* the national markets they were located within, international executive search firms now increasingly stretch the headhunting process across borders through collaboration between international offices using transnational strategies (the shift from a multinational to a transnational form as described by Bartlett and Ghoshal, 1998). The market for executive labor recruitment has changed with the growth in importance of the 'knowledge economy' in developed nations. This intensification of demand for talented individuals to fill executive and boardroom vacancies has created new international geographies of executive labor search. As Tully (1990, 30) wrote '[T]he hunt for the global manager is on. From Amsterdam to Yokohama, recruiters are looking for a new breed of multilingual, multifaceted executive who can map strategy for the whole world. This action is especially heavy in Europe'.

Headhunters have both responded to and driven such change through the stretching of all of the facets described above (for example, functional and industry practice groups as well as through the search process itself) across borders through FDI and office networks. Below, we provide explanations for the growth of international executive search in Europe, highlighting four key drivers of internationalization: demand factors; technological advances; the need to overcome off-limits 'market blockages'; and the creation of a new 'intermediaries' marketplace in executive labor recruitment.

Demand Factors

Organic office growth and merger and acquisition activity spurred the internationalization of executive search firms to Europe and beyond from the USA, and during the 1980s from London, Zurich and Paris to US cities and beyond (Cuthbertson 1996; Jenn 2005; Jones 1989). As discussed earlier, given the bespoke, knowledge intensive and secretive process and practice of headhunting, the most efficient form of serving clients' labor market requirements is through office networks. Quite simply, headhunters have internationalized to serve their clients, who are often transnational firms in all sectors of the economy. The increased demand for externally recruited executive labor and the subsequent scarcity of such individuals have created record levels of demand for the services of headhunters (*The Economist* 1990). As Khurana (2002, 30) points out, the elite labor market is somewhat 'imperfect' and inefficient in an orthodox economics sense where, '[B]uyers and sellers in disconnected and sparse networks may not be aware of the full range of trading partners or opportunities for exchange'. This reinforces the demand for headhunting services to fill the 'structural holes' (Burt 1992) and create the necessary 'weak ties' (Granovetter 1973) between firms and candidates for their vacancy. Lubricating this market requires search with international dimensions both because of the inability to recruit the quantity and type of individual needed through intra-national strategies and because of the global outlook of the multinational firms that are the key clients. The internationalization of executive search is a response to this new dynamic and has intensified since the early 1990s. Especially in Europe, the creation of a single market and the liberalized flow of goods, services and most importantly people since 1992 have reinforced this trend (Britton et al. 1995; Jenn 1993). Internationalized executive search has therefore become a much demanded service, to which executive search firms have responded. For example, it was noted in *The Independent* (2005) that London's legal and accountancy professional service firms were increasingly looking beyond the UK to recruit the talented individuals it needed. Skills shortages have become a real issue, something headhunters can help address through international search strategies.

Technological Advances

Technology has revolutionized the geography of headhunting. The creation of candidate shortlists is a process that draws on researchers and consultants' knowledge of suitable candidates as well as the complex computer databases headhunters now create. Such databases have always existed, previously as card file systems (Jones 1989). However, the internationalization of executive search has been facilitated by the development of databases that, through Internet connections, allow extensive but also intensive searches for potential candidates through the web-portals of major multinational corporations. *Fast Company* (2000, 44) describes the techniques of 'X-raying' and 'Flipping' that use algorithms, run through search engines such as Google, to search a wide range of firms and create lists of candidates for a range of different types of vacancy. These algorithms reveal the contents of non-public webpages that, although legal to access, are not normally displayed or directly accessible to the general public. It is possible, for example, to find listings of executives working on a project and their role by accessing various 'hidden' parts of a firm's website. This can then be used to store the personal details, experiences and skills of individuals who could become candidates in the future. This has dramatically improved the ability of headhunters to search across borders. It is now possible to trawl any organization with an Internet fronted website for suitable talent, regardless of the separation of the potential candidates, their firm and the executive search organization (as discussed by *Fast Company*).

Technology advances have also been important in the internationalization of headhunting because of the integration of offices and the collaboration on global search projects enabled through transnational project architectures (Bartlett and Ghoshal 1998). The use of telecommunications including videoconferencing, globally accessible databases and project intranet sites, allows consultants in several offices to share knowledge of potential candidates. This has become vital as '[T]he number of offices is not as important as the quality of interaction between the consultants...integrated firms function more efficiently for pan-European cross-border assignments' (Jenn 1993).

Self-Regulation and the Need to Overcome 'Off-Limits' Market Blockages

The 'off-limits rule' in headhunting notes that for two years no partner/consultant of an executive search firm may either speak to a person recruited for a previous client or any one else who has worked for a firm who has been a previous client (see for example, Britton et al. 1997; 2000; Konecki 1999). However, the off-limits rule can create a 'blockage problem' as firms find it difficult to seek repeat business with recent successful placed candidates and clients. But, given that the industry is lightly self-regulated by the Association of Executive Search Consultants (AESC) (Konecki 1999), executive search firms have looked for ways to circumvent the 'blockage' problems created when executive search firms work for several different clients in one industry. Internationalization has been used as one strategy to avoid this problem. It means headhunters can draw candidates from a pool of firms spread over a wider geographical area and operate policies that place clients off-limits only to headhunters working in the offices that served the client. The latter is a strategic ploy executive search firms increasingly use because it means that a firm that was a client of one or several offices can be targeted by a number of other offices in the firm's network whilst remaining off-limits only to those offices that directly served the firm as a client (for example see Konecki 1999).

The Creation of a New Intermediary's Marketplace for Executive Labor

Whilst the disintermediation of many KIBS has followed the rise of the internet and e-commerce (for example see Leyshon and Pollard 2000), the executive search industry, like temping (Peck, Theodore and Ward 2005) has firmly established its standing as a vital intermediary over recent years. As was noted earlier, the scarcity of talented individuals has led to recognition of the need to recruit individuals from outside the organization to fill executive positions. However, as sought after candidates are already employed, and often by a competitor firm, it is imperative that searches are covert because of their commercial sensitivity. It is therefore essential that the candidate's current employer does not become aware that their leading executive is about to be headhunted, and that candidates do not know who has employed the headhunted until they are committed to interview for the post. Headhunters are seen as being able to manage this process most effectively, preserving both the client's and the candidate's confidentiality. So, in effect, headhunters can open doors to candidates that would be closed to direct approaches from a firm with an executive vacancy. This 'socially constructed' nature of elite labor search requires the ability to negotiate the embedded nature of acceptable talent hunting (see Granovetter, 1985). Executive search firms have the expertise and ability to do this and successfully connect supply (candidates) with demand (executive vacancies) whilst not crossing or contravening the socially-defined boundaries and norms of headhunting. As intermediaries, they ensure that the interests of both the client and candidates are best served. Britton et al. (2000) described this as a three-way relationship in which the executive search firm, at the centre of the relationship, has to manage the complexities of this embedded market in terms of both candidate and client expectations.

Geographical Expansion of the Worldwide Offices Networks of Europe's Leading Transnational Headhunters

Following on from Boyle et al.'s (1996) analysis of the expansion of European headhunting world city office networks until the early 1990s, in this section of the chapter we chart the global growth of the top sixteen European headhunters from 1992 to 2005, drawn from The Economist Intelligence Unit's report on *Executive Search in Europe* (Jenn 1993) and Jenn's recent (2005) publication, *Headhunters and How to Use Them*, supplemented by individual firm Annual Reports and world-wide web sites.

World Regional Office Change, 1992–2005

A basic analysis of office expansion and distribution indicates that the total number of offices operated by the sixteen largest firms at each census point increased by 55 per cent between 1992 and 2005, from 471 to 729 (Table 7.3). During this time period Europe remained the dominant region for office concentration in both absolute and relative terms, accounting for 261 offices in 1992 and 325 in 2005. When North American (NA) office numbers are added to Europe, both regions accounted for the highest share of total world offices, 79 per cent and 69 per cent in 1992 and 2005, respectively. However, at this regional scale, an analysis of the world-wide office data also reveals two other significant trends in the distribution of office expansion. The first trend is the consolidation and growth of offices in the Asia-Pacific (AP) region, in both absolute and relative terms, from 67 to 142 offices, representing an increase of 112 per cent between 1992 and 2005. The second trend is the growth of office networks in the genuine 'emerging markets' of these leading firms, Latin America (LA), and Africa and the Middle East (AME), with increases of 148 per cent (from 29 to 72 offices) and 1500 per cent (from 3 to 16 offices), respectively between the two time periods.

Transnational Firm Office Change, 1992–2005

At the firm level, where regional office data is comparable between the same top firms in Europe in 1992 and 2005 (see Table 7.4) it is interesting to note that: (a) in absolute terms the highest office expansion was recorded in the Paris and joint London-New York headquartered firms TransSearch International and Ray & Berndtson (both recording +35 offices), but with respect to relative growth, it was Ray & Berndtson with a +233 per cent increase in offices from 15 to 50 which

Table 7.3 Europe's sixteen leading transnational executive search firms: office change by world region, 1992–2005 (ranked by revenue)

Region	1992[1]	Per cent share	2005[2]	Per cent share	Absolute change	Percentage change
Europe	261	55	325	45	+64	+25
North America	111	24	174	24	+63	+57
Latin America	29	6	72	10	+43	+148
Asia-Pacific	67	14	142	19	+75	+112
MiddleEast and Africa	3	1	16	2	+15	+1500
Totals	471		729		+258	+55

Source: [1] Adapted from Jenn (1993); [2] Firm worldwide web accessed 07/12/05.

Table 7.4 Firm world regional office network change 1992–2005 (ranked by 1992 firm)

Firm	Number of offices in each word region															Total change	% change
	Europe			North America			Latin America			Asia-Pacific			Africa / Middle East				
	1992	2005	%Δ	1992	2005	%Δ	1992	2005	%Δ	1992	2005	%Δ	1992	2005	%Δ		
EZ	24	28	+17	7	11	+71	3	6	+100	5	11	+120	0	3	-	+20	+51
AM	27	40	+48	11	9	-18	4	9	+125	7	17	+143	1	3	+200	+28	+56
SS	15	19	+27	12	17	+42	1	5	+400	4	7	+75	0	1	-	+17	+53
H&S	11	22	+100	17	17	-	0	7	-	2	11	+450	0	1	-	+28	+93
BE.	15	28	+87	0	12	-	0	4	-	0	5	-	0	1	-	+35	+233
WH	17	18	+6	10	5	-50	3	1	-67	8	6	-25	0	0	-	-8	-21
RR	7	12	+71	11	12	+9	0	2	-	5	6	+20	0	0	-	+9	+39
KF	12	23	+92	18	23	+28	6	10	+68	7	15	+114	0	2	-	+30	+70
BN	18	28	+56	12	13	+8	3	6	+100	11	17	+55	1	2	+100	+21	+47
Trans.	25	31	+24	1	13	+1200	3	9	+200	3	13	+333	0	1	-	+35	+109

Key: EZ Egon Zhender; AM Amrop (merged in 2000 with Hever to form Amrop Hever Group); SS Spencer Stuart; H&S Heidrick & Struggles Int.; BE. Berndtson International (merged in 1990s with Paul Ray to form Ray & Berndtson Int., and merged with Odgers in 2000); WH Ward Howell (changed their name to Signium in 1998); RR Russell Reynolds; KF Korn/Ferry Int.; BN Boyden Int.; Trans TransSearch.

Source: 1992 data adapted from Jenn (1993) and 2005 data sourced from firm worldwide web sites, accessed 07/12/05.

topped the table between 1992 and 2005; (b) all firms with the exception of Signium (formally known as Ward Howell) had two-digit relative percentage increases in their office networks in Europe, with significant increases by Heidrick & Struggles (+100 per cent to 22 offices); (c) only four firms recorded two-digit relative percentage increases in offices networks in North America (Egon Zhender; Spencer Stuart; Korn/Ferry offices; and TransSearch), and both Amrop Hever and Signium reduced their office networks in the region by -18 and -50 per cents, respectively; (d) the highest relative growth rates were recorded by firms in the LA and AP regions, with firms either extending their already established offices networks in the regions (for example, Egon Zhender +100 per cent to 6 offices and +120 per cent to 11 offices in SA and the AP respectively) or opening new offices for the first time (for example, Heidrick & Struggles opened seven new offices in LA; and Ray & Berndtson opened five new offices in the AP); and (e), new absolute growth in the AME region was muted compared to growth elsewhere, but six firms did open up new offices here for the first time between 1992 and 2005, and two firms who were in the region in 1992 had increased their presence since 2005 by on average one extra office per firm.

World City-Office Change, 1992–2005

Unlike many other KIBs (for example, accounting), Europe's leading headhunters operate globally from a relatively small network of world city offices with usually one office per firm per city. In 1992, Europe's sixteen largest firms had 471 offices distributed in 111 world cities, with Europe dominating this distribution (Table 7.5). By 2005, Europe's sixteen largest firms operated 729 offices in 161 cities, with the highest share of offices in European cities, but the largest relative growth rates in AME (+700 per cent) and the AP (+71 per cent) (Table 7.6). An analysis of the change in concentration and magnitude of offices in cities in each region is shown in Table 7.5 for 1992 and 2005, and four major trends stand out from these different city-offices data sets. First, the major world cities in each region accounted for the highest number of offices in both 1992 and 2005 (but London's position was only ranked as number four in 1992 in Europe), and moreover, these cities experienced further office growth between 1992 and 2005 (London +2 offices to 16; New York +3 office to 16; Sao Paulo (+5 offices to 13; Tokyo +3 offices to 13; and Johannesburg +2 offices to 5). Second, significant new office growth occurred in a number of LA and AP cities which before 1992 had never had offices of Europe's top sixteen firms (for example, Santiago +9 offices; Lima +6 offices; Rio de Janerio +4 offices; Shanghai +9 offices; Beijing +7 offices; New Delhi +5 offices and Dubai/ Wellington each +3 offices). Third, in Europe, offices expanded in many of the post-Communist economies of the East (for example, Warsaw + 8 offices from 1992; Moscow +5 offices; Istanbul +7 offices), and in the USA there was growth in Denver and Miami. Fourth, in the AME region in 2005, Johannesburg and Dubai dominated the concentration of offices with half of the grand total, where the other eight offices were spread between six other cities (Sandton; Jeddah; Tel Aviv, Cape Town, Beirut and Kuwait City (2 offices)) (Tables 7.5 and 7.6).

Table 7.5 City-locations of number of international offices of the sixteen leading European transnational headhunters, 1992 and 2005

Number of offices in each city by world region[1]

Europe	2005	1992	North America	2005	1992	Latin America	2005	1992	Asia-Pacific	2005	1992	Middle East/Africa	2005	1992
London	16	14	New York	16	13	Sao Paulo	13	8	Sydney	13	11	Jo'burg	5	3
Paris	15	16	Chicago	12	8	Buenos Aires	9	4	Tokyo	13	10	Dubai	3	0
Amsterdam	14	9	Toronto	12	8	Santiago	9	0	Hong Kong	12	12			
Frankfurt	14	12	Atlanta	11	9	Bogota	8	2	Singapore	12	7			
Milan	14	16	Los Angeles	9	7	Mexico City	8	7	Seoul	9	3			
Madrid	13	15	Houston	8	6	Caracas	6	4	Shanghai	9	0			
Warsaw	12	4	Miami	8	0	Lima	6	0	Mumbai	8	1			
Munich	12	9	San Francisco	8	6	Rio de Janeiro	4	0	Beijing	7	0			
Brussels	11	13	Boston	7	4				Kuala Lumpur	7	3			
Stockholm	11	9	Dallas	7	7				Melbourne	7	9			
Copenhagen	11	8	Montreal	6	4				Bangkok	6	2			
Vienna	10	7	Washington	6	4				Jakarta	6	1			
Zurich	10	12	Denver	5	0				Auckland	5	2			
Barcelona	9	9	Minn./St. Paul	5	3				New Delhi	5	0			
Helsinki	9	7	Palo Alto	5	3				Taipei	4	4			
Budapest	8	7	Calgary	3	1				Manila	4	1			
Istanbul	8	1	Philadelphia	3	1				Wellington	3	0			
Oslo	8	5	Stamford	3	6									

Düsseldorf	8	11	Menlo Park	3	2
Lisbon	8	6			
Moscow	7	2			
Hamburg	7	4			
Geneva	6	6			
Rome	6	7			
Prague	6	3			
Athens	5	2			
Dublin	5	6			
Gothenburg	4	4			
Lyon	4	2			
Bucharest	3	1			
Berlin	3	6			
Leeds	3	3			
Stuttgart	3	3			
Malmo	3	2			
2005 Sub-totals					
34 cities (287 offices)	19 cities (137 offices)	8 cities (63 offices)	17 cities (130 offices)	2 cities (8 offices)	
1992 Sub-totals					
28 cities (231 offices)	14 cities (88 offices)	4 cities (23 offices)	8 cities (82 offices)	1 city (3 offices)	
2005 Grand Totals					
67 cities (325 offices)	48 cities (174 offices)	14 cities (72 offices)	24 cities (142 offices)	8 cities (16 offices)	
1992 Grand Totals					
54 cities (261 offices)	33 cities (111 offices)	9 cities (29 offices)	14 cities (67 offices)	1 city (3 offices)	

Note: [1] Cities with three or more offices.

Sources: firm worldwide websites, accessed 06/12/05 (see appendix 1), adapted from Jenn (1993).

Case Studies

The following case studies are constructed to emphasize the nuances of internationalization and organizational form in three firms: Egon Zehnder, The Globe Search Group; and Amrop Hever.

Egon Zehnder International (Wholly-Owned Firm)

Egon Zehnder International, with headquarters in New York and Europe, is a wholly owned headhunting firm which was the very first firm to be set up outside of the USA in 1964, by its founder of the same name who had worked for Spencer Stuart (Jenn, 2005; www.egonzehnder.com). After the opening of its first office in Zurich in 1964, its first years of internationalization were focused on consolidation in Europe, with new starts in Paris (1968), Brussels (1968), Copenhagen (1970) and London (1970). This was followed by step-by-step internationalization in Asia and Latin America (Tokyo – 1971, the first office to be opened outside of Europe; Sao Paulo – 1975; Mexico City – 1982; Buenos Aires – 1984). Egon Zender did not enter the USA until 1977 with the opening of the New York office, followed by Chicago in 1981 (it opened its last office in the USA in 2000 – Miami). Expansion into eastern Europe came in the early 1990s (for example, Budapest and Prague in 1992), followed by expansion in the other emerging market regions (www.egonzehnder.com). The firm's mission statement is to specialize 'in assessing and recruiting business leaders with outstanding track records who will create competitive advantage and sustainable value' (Egon Zehnder, 2001). As a wholly-owned firm, it is managed by a partnership that defines priorities and strategies for all 'branch' offices. The firm uses its wholly-owned organization as a competitive advantage to demonstrate global-local knowledge structures wherever clients may be in the world:

> The most fundamental expression of our client-first vision resides in our structure, which is unique to our profession. Our 300 consultants, operating from more than 55 wholly owned offices in over 33 countries, are organized around a single-profit center partnership. This is designed to eliminate competitive barriers between our offices. It allows us to operate seamlessly when engagements call of us to mobilize across many offices in a country or a region (www.egonzehnder.com, accessed 07/12/05).

Egon Zehnder uses its wholly-owned structure as a mechanism to nurture long-term, sustainable relationships with clients so that whenever they require the services of a headhunter, they turn to one particular firm, their 'retained' firm, rather than accepting business from other competitors. Egon Zehnder specializes in several global practice groups: financial services, consumer, life sciences, technology and telecoms, energy and process, engineering and automobile, services, private capital and e-commerce The firm uses its extensive office-network located around the world (Figure 7.1) to to reproduce a highly-integrated transnational business model, offering all clients the same branding and corporate culture. In 2004, Egon Zehnder was the world's fourth largest headhunter ranked by worldwide fee income (Table 7.1).

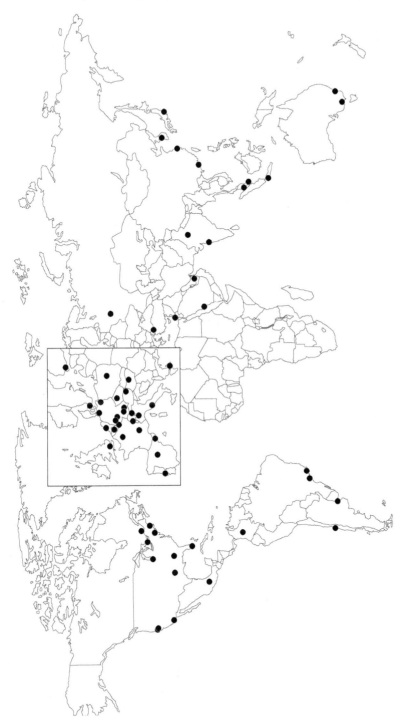

Figure 7.1　The worldwide office network of Egon Zehnder, 2005

Table 7.6 Europe's sixteen leading transnational headhunting firms world regional office networks, 2005 (ranked by revenue)

Firm (founded)	Number of offices in each world region					Total
	Europe	*North America*	*Latin America*	*Asia-Pacific*	*Middle East/Africa*	
1. Korn/Ferry International	23	23	10	15	2	73
2. Heidrick & Struggles Int'l	22	17	7	11	1	58
3. Ray & Berndtson	28	12	4	5	1	50
4. Russell Reynolds Ass.	12	12	2	6	0	32
5. Spencer Stuart	19	17	5	7	1	49
6. Amrop-Hever	40	9	9	17	3	78
7. Whitehead Mann Group	7	2	0	1	0	10
8. Globe	10	4	0	1	0	15
9. IIC Partners	26	12	3	10	2	53
10. TransSearch	31	13	9	13	1	67
11. Signium Int.	18	5	1	6	0	30
12. Eric Salmon & Partners	4	1	0	0	0	5
13. A.T.Kearney	12	11	1	4	0	28
14. Stanton Chase Int.	17	12	9	18	0	56
15. Egon Zehnder Int'l	28	11	6	11	3	59
16. Boyden	28	13	6	17	1	66
Totals	**325**	**174**	**72**	**142**	**16**	**729**

Source: Firm worldwide web sites, accessed 07/12/05.

Table 7.7 The Globe Search Group network of independent firms in 2005

Associate Firm	Founded	Regional Office	Other Offices	Partners/ Consultants
Alumni AB	1990	Stockholm	Malmo	23
Bjerke & Luther AS	1986	Oslo		NA
Corinthe	1988	Oegstgeest		8[1]
DK Executive Search	1997	Warsaw		26
E L Shore & Associates	1968	Toronto		NA
Geddes Parker & Partners	1989	Sydney		5[1]
GlobeSearch A/S	1989	Copenhagen		2[1]
Hofman & Heads!	2003	Munich	Frankfurt	11
Herbert Mines Associates	1981	New York		8[1]
Ment Associates MV	2001	Antwerp		2[1]
Rusher, Loscavio & LoPresto	1977	San Francisco	Palo Alto	6[1]
Sainty, Hird & Partners	1996	London		4[1]
The Miles Partnership	NA	London		5[1]

NA Not available; [1]partners only.

Source: www.globesearchgroup.com/locations.htm accessed 13/01/06.

The Globe Search Group (Network Firm)

The Globe Search Group, with headquarters in New York, London and Munich, is a network organization of 13 independent firms formed in 1997 by Miles Broadbent in London (Table 7.7), operating primarily in Europe, North America and the Asian-Pacific (Figure 7.2; Table 7.6). The network model of organization in the Global Search Group combines non-transnational, local firms in a 'best friend' strategic alliance to meet clients' executive labor search needs worldwide. Global Search is not a single entity like Egon Zehnder or other wholly-owned firms. Each firm performs labor searches independently and calls upon other members of the network when it requires an overseas search in a member's national market. Each independent firm associated with Globe's network cultivates an image of being 'local' to the market in order to maintain 'a very tightly managed client list and seeks to build deep and lasting relationships with its clients' (www.globesearch.com, accessed 13/01/06). Many member firms have been established for fifteen years or more. Thus, several members of Globe's network may operate in one country, competing for the same business outside of the 'off-limits' mantra. Hence, the Globe Search Group can operate effectively as an international collection of independent firms 'by combining international reach with regional excellence, the Globe Search Group expertly supplies the resources, experience and knowledge in the identification of key executives internationally' (www.globesearchgroup.

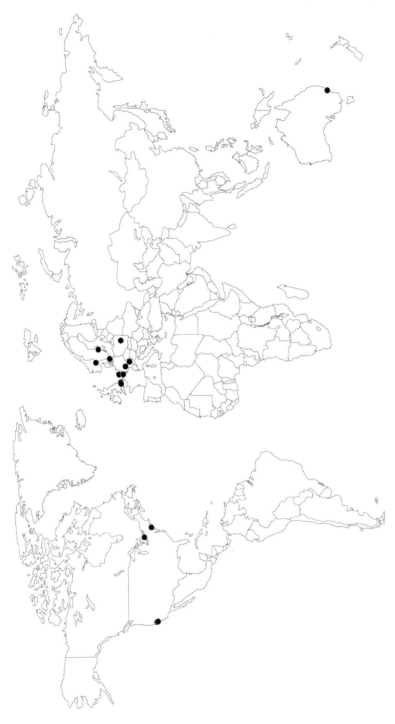

Figure 7.2   The worldwide office network of the Globe Search Group, 2005

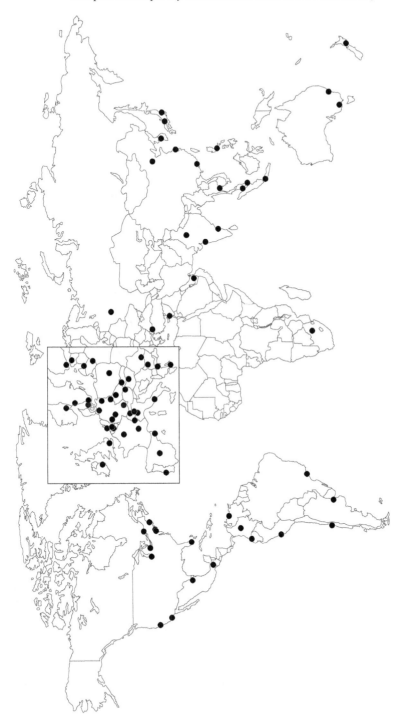

Figure 7.3   The worldwide office network of Amrop Hever, 2005

com/, accessed 13/01.06). As such, the Globe Search Group represents a corporate reincarnation of the 'temporary teams' Grabher (2002) identified within advertising agencies, which are *temporary* and *dynamic* inter-firm coalitions that grow through repetitive use, and are very ephemeral in nature. In 2004, the Globe Search Group was the eighth ranked firm in Europe and tenth worldwide (by fee income) (see Table 7.1).

The Amrop Hever Group (Hybrid Firm)

The Amrop Hever Group is a hybrid transnational headhunter, a formal global partnership of 59 independent firms that trade under one business model (Jenn, 2005; www.amrophever.com) with a very strong global identity. Member firms often incorporate Amrop and/or Hever into their names, but also retain their local names in their markets. The Group is managed by a Chairman, three Vice-Chairmen (one each from Europe/Middle East & Africa, the Americas and Asia-Pacific) and a seven-member Board of Directors (from UK, Canada, Japan, India, China, Sweden, USA). Amrop Hever is the largest firm in Europe, with 70 offices in over 50 countries (Table 7.1; Figure 7.3). The Group was established in 2000 through the merger of Amrop International and the Hever Group 'to serve more efficiently regional and international clients in light of the growing trend of globalization ... speak[ing] over 100 languages but always with a single voice' (www.amrop.com/aboutus.html, accessed 12/06/05). Amrop Hever is more tightly-integrated than The Globe Search Group with an overall global headquarters in Brussels where the Worldwide Secretariat formulates strategy and policy, but it is not as culturally, socially or economically incorporated as the wholly-owned firm, Egon Zehnder.

Discussion and Conclusions

Our contemporary analyses of the leading European transnational headhunting firms and practices have served to deepen an understanding of both the complexities of internationalization and the strategies by which labor-related KIBS intermediate in the marketplace. The entire rationale for internationalization is client-focused, which spills over into a classic herding mentality as leading firms follow competitors into international markets to secure market share and maximize profitability. We offer four observations to synthesize the ways in which the internationalization of the leading European transnational headhunters reproduce their intermediation functions in the recruitment of executive labor at all geographical scales.

First, we have noted that the internationalization process is not uniform across firms with respect to the mode and strategy of internationalization. Firms have internationalized through FDI and wholly-owned growth, as in Egon Zehnder for example. Alternatively, penetration into new markets has been in the form of networks of independent firms (for example, The Globe Search Group) and hybrid forms that combine tightly organized and independent firms (Amrop Hever). Network and hybrid organizational forms in particular have been used by different firms at particular moments in their institutional history to spread risk in new markets through minimising the sunk costs of new starts (i.e. wholly-owned starts). The hybrid organizational form (and to a lesser extent the network) is also a vehicle for gaining immediate reputation in a new location by aligning with knowledge rich, local firms with significant experience in intermediation in the local executive labor market.

Second, and linked to the first, internationalization through FDI has been reproduced by the requirement of firms to overcome the 'off-limits' best practice. By minimising these 'Chinese walls' between independent firms (as in a network or hybrid form) and wholly-owned international offices, firms have been able to undertake transnational searches to overcome self-imposed regulatory frameworks to recruit executive labor from recent placements and clients.

Third, it is evident that the operation of the executive search industry in Europe and beyond is a classic manifestation and reproduction of the 'Americanization' of corporate services in the world economy (see for example, Daniels, 1999). The Big Four US firms set new standards and benchmarks in the operation of executive search, recreating deep-rooted intermediary relationships with clients as 'market makers'. These practices very quickly became the norm in labor markets of all scales as local firms had to adapt to competition.

Fourth, the pattern of internationalization as illustrated in worldwide office change for the sixteen leading firms in Europe clearly shows the desire to establish new ventures in emerging markets (like LA, AP and AME) with respect to all three forms of internationalization strategy. At the same time, the industry remains consolidated in Europe and the USA. Europe remains the leading market for the industry as indicated by fee income (Table 7.1) and office concentrations in 2005 (Tables 7.3 to 7.6).

To conclude, we reiterate that this case study of European transnational headhunters has been used as an exemplar to enhance the knowledge of the internationalization of KIBS. The industry is dominated by the operations and practices of transnational corporations, tied to world-city markets, but like many other KIBS, these are supported by thousands of SMEs, who specialize in narrow segments of the labor market, nearly impossible to quantify. Headhunting internationalization strategies are client-focused, but simultaneously are devised to make their own markets with clients and to continually nurture their intermediary role. The future of headhunting is firmly established in the physique of the intermediary executive labor market as clients will always wish to recruit 'stars' employed by competitors, and individuals will always be willing to be seduced by the approach of headhunters who add significant gravitas to any new employment opportunity.

Acknowledgements

We would like to thank the United Kingdom's Economic and Social Research Council for funding this research in the project, 'The globalization of the executive search industry in Europe' (Award Number: RES-000-22-1498).

References

Bartlett, C. and Ghoshal, S. (1998), *Managing Across Borders: The Transnational Solution*, 3rd ed. (London: Random House).

Beaverstock, J. (2006), 'Connecting World Cities: Organizational Labor Mobility in Transnational Banking', in Hammen, Joost, Koekoek, Arjen, Velema, Thijs and Verbeet, Mark (eds), *Be Connected. Global Cities and The Space of Flows* (Utrecht: NGS).

Boyle, M., Findlay, A., Lelievre, E. and Paddison, R. (1996), 'World Cities and the Limits to Global Control: A Case Study of Executive Search Firms in Europe's Leading Cities', *International Journal of Urban and Regional Research* 20: 3, 498–517.

Britton, L., Doherty, C. and Ball, D. (1995), 'Executive Search and Selection in France, Germany and The UK', *Zeitschrift Fur Betriebswirtschaft* 67: 2, 219–232.

Britton, L., Doherty, C. and Ball, D. (1995), 'Alliances in Executive Search and Selection', in Curwen, Peter (ed.), *The Changing Global Environment: Case Studies* (Sheffield: Sheffield Hallam University Business School, Policy Research Centre), 8–15.

Britton, L., Wright, M. and Ball, D. (2000), 'The Use of Co-Ordination Theory To Improve Search Quality in Executive Search', *The Service Industries Journal* 20: 4, 85–102.

Bryne, J.A. (1986), *The Headhunters* (New York: Macmillan).

Bryson, J., Daniels, P.W. and Warf, B. (2004), *Service Worlds* (London: Routledge).

Burt, R. (1992), *Structural Holes: The Social Structure of Competition* (Cambridge: Cambridge University Press).

Coe, N. (1997), 'Internationalization, Diversification and Spatial Restructuring in Transnational Service Firms: Case Studies from The UK Market', *Geoforum* 28: 3–4, 253–270.

Cuthbertson, N. (1996), 'The Geography of The International Headhunting Industry', Unpublished Phd Thesis, University of Bristol.

Daniels, J.D., Radebaugh, L.H. and Sullivan, D.P. (2004), *International Business. Environments and Operations*, 10th ed. (New York: Prentice Hall).

Daniels, P.W. (1999), 'Overseas Investments By US Service Enterprises', in Slater, David and Taylor, Peter (eds) *The American Century* (Oxford: Blackwell), 67–83.

Department of Trade and Industry (1998), 'Our Competitive Future Building The Knowledge Driven Economy', *Department of Trade and Industry White Paper* (Www.Dti.Gov.Uk/Comp/Competitive) (London: Whitehall).

Dicken, P. (2003), *Global Shift*, 4th ed. (London: Sage).

Dunning, J. H. (1993), *The Globalization of Business: The Challenge of The 1990s,* (London: Routledge).

Dunning, J. and Norman, G. (1983), 'The Theory of Multinational Enterprise: An Application of Multinational Office Location', *Environment and Planning A* 15: 675–692.

Dunning, J. and Norman, G. (1987), 'The Location Choice of International Companies', *Environment and Planning A* 19 (5), 613–632.

The Economist (1990), 'Headhunting For the Moneymen. in Search of Excellence', 6th January, 78–79.

Egon Zehnder (2001), *Unleashing the Power of Business Leadership* (Geneva: Egon Zehnder International, Neidhart & Schon AG).

Executive Grapevine (2004), *Directory of Executive Recruitment International Edition* (London: Executive Grapevine).

Fast Company (2000), 'The Great Talent Caper', 38, September, 44.

Finlay, W. and Coverdill, J. (1999), 'The Search Game: Organizational Conflicts and the Use of Headhunters', *The Sociological Quarterly*, 40: 1, 11–30.

Finlay, W. and Coverdill, J. (2000), 'Risk, Opportunism and Structural Holes: How Headhunters Manage Clients and Earn Fees', *Work and Occupations*, 27: 3, 377–405.

Finlay, W. and Coverdill, J. (2002), *Headhunters: Matchmaking in The Labor Market* (Ithaca, NY: ILR Press).

Grabher, G. (2002), 'The Project Ecology of Advertising: Tasks, Talents and Teams', *Regional Studies* 36: 3, 245–263.

Granovetter, M. (1973), 'The Strength of Weak Ties', *American Journal of Sociology,* 78: 6, 1360–1380.

Granovetter, M. (1985), 'Economic Action and Social Structure: The Problem of Embeddedness', *American Journal of Sociology* 91: 3, 481–510.

Gurney, D.W. (2000), *Headhunters Revealed! Career Secrets For Choosing and Using Professional Recruiters* (New York: Hunters Arts Publishing).

The Independent (2005), 'Labor Shortages Spread To Law and Accountancy', Business News, March 1st, P1.

Jenn, N. (1993), *Executive Search in Europe* (London: The Economist Intelligence Unit).

Jenn, N. (2005), *Headhunters and How To Use Them* (London: The Economist Publications).

Jones, S. (1989), *The Headhunting Business* (Basingstoke: Macmillan).

Khurana, R. (2002), *Searching For A Corporate Savior. The Irrational Quest For Charismatic CEOs* (Princeton: Princeton University Press).

Konecki, K. (1999), 'The Moral Aspects of Headhunting. The Analysis of Work by Executive Search Companies in "Competition Valley"', *Polish Sociological Review,* 4: 128, 553–568.

Leadbeater, C. (1999), *Living On Thin Air: The New Economy* (London: Viking).

Leyshon, A. and Pollard, J. (2000), 'Geographies of Industrial Convergence: The Case of Retail Banking', *Transactions of The Institute of British Geographers*, NS 25: 2, 203–220.

Morgan, G. and Quack, S. (2005a), 'Internationalization and Capability Building in Professional Service Firms', in Morgan, G., Whitley, R. and Moen, E. (eds), *Changing Capitalisms? Internationalization, Institutional Change and Systems of Economic Organization* (Oxford: Oxford University Press), 277–311.

Morgan, G. and Quack, S. (2005b), 'The Internationalization of UK and German Law Firms', Organizational Studies, 26: 12, 1765–1786.

Nachum, L. (1999), *The Origins of The International Competitiveness of Firms*, (Cheltenham: Edward Elgar).

Peck, J. and Theodore, N. (2001), 'Contingent Chicago: Restructuring the Spaces of Temporary Labor', *International Journal of Urban and Regional Research* 25: 3, 471–496.

Peck, J., Theodore, N. and Ward, K. (2005), 'Constructing Markets for Temporary Labor: Employment Liberalization and the Internationalization of the Staffing Industry', *Global Networks* 5: 1, 3–26.

Roberts, J. (1999), 'The Internationalization of Business Service Firms: A Stages Approach', *The Service Industries Journal* 19: 4, 68–88.

Sennett, R. (1998), *The Corrosion of Character: Personal Consequences of Work in The New Capitalism* (New York: W.W. Norton & Co.).

Simms, J. (2003), 'Unnatural Selection', *Director* 56, 70–74.

Tully, S. (1990), 'The Hunt for the Global Manager', *Fortune* May 31st, 30–36.

Ward, K. (2004), 'Going Global? Internationalization and Diversification in the Temporary Staffing Industry', *Journal of Economic Geography* 4: 2, 251–273.

Chapter 8

Internationalization of Japanese Professional Business Service Firms: Dynamics of Competitiveness through Urban Localization in Southeast Asia

Patrik Ström

The service sector has risen to become the most dominant part of the advanced OECD countries' economies. One of the most dynamic and important parts of the service economy, and a forerunner in the interconnectedness of global economies, is the professional business service sector. This includes various forms of high-value adding services with high knowledge content, such as information technology consulting, financial related consulting and management consulting. These services have established themselves as important drivers of productivity and efficiency throughout the economy, but they also create a demand for professional services among themselves.

In a comparison with the most advanced OECD countries, Japan appears to be somewhat different in terms of the GDP and employment share generated by the service sector (OECD 2004). Disaggregating the statistics further into the high-knowledge or professional services generates a similar picture. Additionally, Japanese companies have been relatively unsuccessful in competition with Western professional business service firms on the global market (Enderwick 1990; Johansson 1990; Nachum 1999; Ono 2001; Ström 2004; 2005). This perceived lack of competitiveness together with a less developed domestic service economy raises the question of the difficulties facing Japanese professional business service firms. Little research has been conducted on the operations of Japanese professional business service firms both in Japan and abroad. Researchers have called for improved understanding of the service economy in Asia (e.g. Daniels 1998; 2001) through various case study oriented approaches. The focus in this chapter is on Japanese firms operating under international competition in Singapore, a geographical area where these firms should hold competitive advantages based on historical, agglomeration and experience bases.

This research uses primary data obtained from Japanese companies to answer why Japanese professional business service firms seem to have difficulties in international competition and how they are utilizing Singapore as a business location. Furthermore, what measures do the firms take to upgrade their regional competitiveness? These questions are operationalized through the study of spatial and managerial issues in the choice of Singapore as a strategic base in Southeast Asia. The combination of sector- and location specific characteristics helps to explain how these firms have internationalized and continuously try to expand their operations. Since the paper is primarily of empirical character, only a brief theoretical frame and literature review is presented. The aim is not to test theory, but rather to let it be the frame for the empirical research. At a later stage, this research can be used in theory building based in the Asian empirical context, as suggested by Yeung and Lin (2003). The rest of the paper is structured in the following way. After the theoretical frame is presented, the methodological aspects of the research are discussed. After presenting the results from the empirical study and the introduction of three mini-cases, the paper ends with a concluding discussion and issues for further research.

Theoretical Framework

One of the best suited theoretical frameworks for catching complex interrelations between company structure and location is the OLI paradigm (e.g. Dunning 1988). Consisting of ownership, location and internationalization specific advantages, this generic approach towards internationalization generates a theoretical foundation for exploring the internationalization of the Japanese professional business service sector. The eclectic paradigm has also been applied to the service sector (Dunning 1989). Regarding the O-advantages or the competitive advantages, according to Dunning there are six factors which are of importance in the field of services, especially intermediate services. These are: quality and product differentiation, economies of scope, economies of scale and specialization, technology and information, knowledge, and finally favoured access to inputs or markets. They are all important for service firms, because they influence their abilities on the market. Regarding the L-advantages, Dunning gives an extensive presentation of why the location aspects are of such importance for the service oriented TNCs. The main reasons behind the development are changing regulatory patterns and the advancement of technology. Additionally, from the firm perspective, the closeness to customers is the most important factor for locating producer or business services via FDI. The importance of market-size and character is not surprising. In dealing with the so-called I-advantages, transaction costs are vital for understanding the underlying reasons for keeping activities in-house.

For many business services such as financing, IT and organization consultancy, knowledge is tacit and expensive to produce, but can easily be replicated. This is the reason why many service firms keep their operations in-house. Additionally, there are many arbitrary gains to geographically spread in-house activities. A second group of intermediate business service firms that often want to keep activities in-house are firms wanting to ensure productive efficiency together with the quality of the end product. In this group, one finds companies with strong brand-names located in sectors where brand-name and experience matters. Management consultants, advertising companies, and construction firms exemplify these services. Another group of companies that maintain in-house activities are trade-related firms. Their mission is to obtain inputs or market knowledge for parent companies or closely connected customers. Dunning (1998) was one of the forerunners in emphasizing location within international business research. These location-bound assets are becoming more important. Clusters with special characteristics will be of great importance, together with governmentally created locational advantages. This discussion brings another perspective to IB research and the OLI-configuration, because it states that 'the locational configuration of a firm's activities may itself be an O-specific advantage, as well as affect the modality by which it augments, or exploits its existing O advantages' (Dunning 1998, 12).

This research has an advantage of combining the managerial aspects of internationalization and localization, with the more descriptively or conceptually oriented research found in economic geography. Dunning further notes the range of governmental influence, which means that a business-oriented approach is necessary to create and sustain innovative clusters, which can both stimulate the growth of new companies, and also attract existing MNCs. In this way, location-bound aspects are important on the micro level for the firm, but also on the macro level for states and their possibilities to sustain the welfare of its citizens. This discussion is of great importance in the field of economic geography research due to the changing landscape of industrial production and where high-knowledge content services are put forward as new engines of regional development (e.g. Juleff-Tranter 1996; Wernerheim and Sharpe 2003).

The locational dimension for producer services has been discussed, both from a narrow regional perspective (e.g. Sjøholt 1993; Beyers and Lindahl 1996; Hermelin 1997; Keeble and Nachum 2002) and from a wider perspective (O'Farrell et al. 1996), where the regional perspective is related to the internationalization process of firms. Keeble and Nachum (2002) show that professional business service firms are highly dependent on integrated clusters, with high levels of collaboration, knowledge transfer and labour mobility. Furthermore, their study shows that service firms in the vicinity of good transport infrastructure can often extend their client area. Bryson (2000) calls the increased need for knowledge diffusion, the availability of expertise and information, *relationship capitalism*. In combination with reputation and image, this creates much of the foundation for localization for business services (Amin 1999). Company location tends toward industry-specific clusters to communicate the fact that they are part of an innovative, respected and image-creating geographical area.

Method

The study was conducted in Singapore during February 2004, and used in-depth interviews with CEOs of 11 companies, based on a questionnaire that was sent to the respondents two weeks in advance. The methodological approach as been modelled around the concept of 'close dialogue' involving industry representatives, to obtain a deeper understanding of the current business situation (Clark 1998). Additionally, the aspect of actor networks as an extension of the 'close dialogue' (Yeung 2003), has guided the research in an attempt to catch the complex social and business networks existing among Japanese companies.

There are a number of reasons for choosing Singapore as the place of research. First, it is a place where Japanese firms have a long tradition of business activity and foreign direct investment. Many firms have been active here for several decades and especially after the sharp appreciation of the Yen in the mid 1980s, there was a steady inflow of Japanese capital. Secondly, Singapore is one of the most developed and advanced economies in Southeast Asia, acting as a major hub and regional financial centre. This implies that there is a good basis for the existence of professional business service firms. Thirdly, the usage of English in the Singapore business environment was considered to be of help when conducting the interviews. All respondents spoke good English. The interviews had been set up through the help of the Japan Center for International Finance in Singapore.

The questionnaire covered five different areas including general information of the company, *keiretsu* (horizontal industry group) affiliation, issues on location and strategy, customers and competitors. The service sector is characterized by highly diverse operations. In order to narrow the scope of the study, Japanese professional business service firms were chosen. This service sub-sector involves companies in which a university degree often is required to practice. The study consisted of two Research Institute companies (consulting operations in various fields), two insurance companies, two information technology consulting firms, three financial services firms and organizations, one *sogo shosha* (trading company) and finally one advertising and PR company. The companies are all well-established in their respective fields and are among the leading Japanese firms in their particular sub-sectors. Furthermore, the respondent companies have a long tradition of working overseas not only in Asia, but also with offices in North America and Europe. One of the directors was additionally the president of the Japanese Chamber of Commerce and Industry in Singapore, and could therefore refer to a broader perspective of Japanese related business in Singapore. Apart from the company interviews the research project was complemented with an interview with the director of economic research at Japan External Trade Organization (JETRO) in Singapore.

Brief Overview – Japanese Operations in Singapore

Singapore has been an important location for Japanese companies for several decades. Due to the rapid appreciation of the Yen in the late 1980s, many firms chose to locate both service and manufacturing operations to Singapore. Table 8.1 shows the situation regarding outward investment between Japan and Singapore. It shows that Singapore has been an important host country of Japanese FDI since the

Table 8.1 Japanese FDI in Singapore by cases, value* and share of Asian FDI

	1989	1990	1991	1992	1993	1994	1995	1996	1997	1998	1999	2000	2001	2002
Cases	181	139	103	103	97	69	94	102	96	58	51	25	31	34
Value	257.3	123.2	83.7	87.5	73.5	110.1	114.3	125.6	223.8	83.2	110.2	50.5	143.3	91.5
Share: Cases %	10.6	9.3	8.1	8.1	6.6	5.3	5.8	8.3	8.3	10.6	9.6	5.4	6.2	6.4
Share: Value %	23.4	11.9	10.3	10.5	9.6	10.9	9.6	9.6	15	9.8	13.7	7.6	17.8	13.5

Source: Japan Ministry of Finance, 2004. * JPY Billion.

Table 8.2 Distribution of Japanese high knowledge business service firms in Singapore

	1960	61–65	66–70	71–75	76–80	81–85	86–90	91–95	96–00	01–05	Total
Trading Firm	6	..	2	9	16	9	25	24	10	1	102
Finance	1	1		2	1	4	4	4	10	3	30
Insurance	1	1			4	1	3	1	1		11
Investment						1	2	1	1	1	6
Construction	1	1	2	11	15	8	6	1	2		47
Telecom.							2	1	4		7
Business Services	3			1	3	6	6	5	3	1	28
Other services	2	1		1	4	2	4	6	4		24
TOTAL											**255**

Source: Membership Directory 02/03, JCCI Singapore.

late 1980s when Japan experienced the bubble economy and strong outward flow of investment. The impact of the Asian crisis is also visible in the sharp downturn of 1997–98. Despite of losing out primarily to China and other low cost countries in Asia, Singapore has been able to hold its position in terms of value of FDI. The underlying reasons for locating in Singapore will be discussed further below.

Apart from the macro economic data given above, Table 8.2 shows the presence of Japanese high knowledge business service or professional business service firms in Singapore based on the directory of the Japanese Chamber of Commerce and Industry, JCCI. It shows the development over the last 40 years. Due to the classification problems existing in services a rather broad perspective is offered, rather than only showing some of the more narrow service sectors. An example of this is the *sogo shosha*, with activities spanning over a great number of service sub-sectors. Similar problems exist in the construction industry. Engineering consultants are sometimes found within the construction classification, which makes it difficult to single them out. Additionally, several construction companies might not have any building projects running, but rather various forms of consultancy assignments. By including these sectors the intention is to show the broad presence of Japanese firms in Singapore. The total number of Japanese companies active in Singapore was 763 in 2002, including both manufacturing and non-manufacturing. The high-knowledge service sectors constitutes about one third of the total companies active in Singapore.

The Japanese professional business service firms in Singapore are fairly limited in size. More than half of the companies in the respective category in the table above have less than 50 employees. The only exception is the construction industry, with several firms employing more than 100 persons. Compared with many other Japanese manufacturing firms working abroad the size of these firms might seem small. In the service sector, however, it is very common that firms are small and this is no particular characteristic of Japanese business service firms. The same situation has earlier been found among Japanese firms active on the UK market (Ström 2003).

Results of the Empirical Study

Who Are They?

Apart from the *sogo shosha*, one financial institution and one of the information technology companies, all companies sat up their business during the 1970s or early 1980s. This shows the large interest in this location even before the sudden appreciation of the yen in 1985. The Japanese companies perceived Singapore to be an important location for various kinds of service functions, often in connection to the expansion of Japanese manufacturing firms into the area taking advantage of the suitable cost structure and the potential market. The overall reason for the Japanese companies to come to Singapore was the large number of Japanese clients already present. This enabled the Japanese companies a solid ground for their businesses, and shows strong similarities with how Western service companies have followed clients abroad, and at later stages expanded their business operations (Sharma 1991; Eramilli and Rao 1993; Majkgård and Sharma 1998). What has been the outcome of this internationalization? Many Western firms seem to have developed their foreign market-oriented operations on a learning and commitment basis (Johanson and Vahlne 1977; 2003). This means that the company becomes more committed to the international market with increased experience. Incremental expansion of the client base is often the result. Judging from the market situation, the ability to use experiences and strengthening competitive advantages have been limited for Japanese firms. We will return to this issue. In terms of employees the Japanese companies' Singapore operations can be characterized as small- or medium-sized, even though the overall firms are some of the biggest in their respective fields. In the interview sample, two firms have more than 250 employees, and two are smaller with about 10–15 employees, with the rest of the companies in between.

The main business activities of the interviewed companies cover many sub-sectors of the professional business service industry. The insurance companies are the only example where it is possible to speak of one, very distinct focus. The operations in Singapore exclusively cover liability non-life insurance, with no interests in the life insurance sector. This focus reflects the difficulties of operating and getting licenses in a foreign market. Apart from the pure non-life insurance activities, the information gathering function for the mother company in Tokyo should not be neglected. The other companies that were interviewed are broad service suppliers in their respective areas. Their operations in Singapore are often more streamlined than the operations in the home market. In the case of the research institute companies they are more specialized in running financial related consultancy and research than in Tokyo where think tank operations and broader business consultancy work is more common. The *sogo shosha* that was included in the study was one of the few companies that covered most of the operations that it undertakes in its domestic market.

Does the Keiretsu Affiliation Matter?

The *keiretsu* structure is very complex, involving relations on both the horizontal and vertical levels (Gerlach 1992). This makes it complicated to determine how firms are connected to each other and if they are equal partners or sub-suppliers. In the case of professional business services, however, it is mainly the horizontal relations that are important. The connections that exist within a group in terms of services often concern a flow of information, not necessarily entailing a control mechanism. Secondly, a *keiretsu* structure of related companies assisting and owning each other can also exist within a larger single company structure. Due to its size, the Toyota group resembles a *keiretsu* structure, with a large number of vertically and horizontally connected firms (Ström 2004). This means that large company groups can work as important areas of business creation and control for business service providers. The prolonged economic problems of the Japanese economy and fierce competition in all international markets, along with the increased globalization of capital markets and supply chains have forced the traditional *keiretsu* system to change (McGuire and Dow, 2003; Ström and Mattsson, 2005). The pressure of financial results has forced many of the Japanese companies to apply a more short term view on international business operations. Relationships that were important in the early stages of the internationalization, and the contact network among company or *keiretsu* group firms that have been utilized for contracts do not have the same strength today. Financial results are now more important than keeping relations intact. To say that the *keiretsu* structure is dissolving is not true, rather it is changing and adapting to a new business environment.

Four of the companies interviewed do not belong to any specific *keiretsu* group, but are all well connected in a larger company group structure. This fact often means that they are more dependent on the mother company. They can for example be working on the international market only to support the mother company in Japan and have limited flexibility to make their own strategy and choose their own clients. The other companies who are members of a larger *keiretsu* emphasized the importance of group support in the early internationalization phase. These firms also have a large share of revenue from *keiretsu* companies. Despite their relations and good contacts with other firms in the *keiretsu*, these firms rarely have more than 30 per cent of their total revenue from the *keiretsu* group. The result further strengthens the perception of a pull factor in internationalization of professional business service firms. All of the interviewed firms state that the main reason for entering the Singapore market was to supply services to the mother company or other firms found within the same *keiretsu* group. Over time the situation has then changed, and the lion's share of the business at present mainly comes from clients outside the single mother company group or *keiretsu*. The data also supports the ongoing changes in the *keiretsu* structure discussed above. The firms all feel the pressure to extend their customer base, and that earlier *keiretsu* connections have lesser value in competition for contracts.

Why Singapore?

In an ever increasing competitive world market, the choice of geographical location and the utilization of this in the overall strategy have become paramount (Porter 2000). Information technology has enabled service firms to offer their services from long distances, but these firms seem also very dependent upon a specific 'atmosphere', where it is important to have a foothold to exist in the industry. Additionally, close proximity to clients, competitors and regulators generates a fruitful ground for business development. This new form of geographical proximity has been widely discussed in the concept of 'being there' (Gertler 2003) or a shift towards a more relational economy (Bathelt and Glückler 2003). The study at hand has tried to explore this aspect based on a two-pronged approach, asking companies' executives why they choose Singapore as a business location and how they are working to utilize the geographical location in the broader strategy of their companies.

The majority of the Japanese companies use the Singapore office as a regional headquarters. However, there are big differences among the firms in the level of authority of these regional functions. It is most common that the Singapore office has an administrative supportive role for branch offices located in nearby countries. Since many of these offices are small, the role of the operation in Singapore is also to provide product line or specialist support when needed. Nevertheless, many of the firms describe a situation where the company head office in Tokyo exerts power both in terms of regional planning on the administrative level, but even more so in the case of product specialists. The respective product line heads and the head of the international department are located in Tokyo. This generates a dual structure of command where the regional headquarters are not as independent as they could be if they were acting as a true regional authority.

Additionally, the geographical scope of the regional headquarters differs widely. In some cases the regional headquarters comprises Singapore and its closest neighbours, the ASEAN countries, but there are also firms that include the Middle East and Australia. This gives a very broad geographical spread that can be difficult to manage in an efficient way. What goes on in the North automatically have implications on the business opportunities in the Southeast Asian region, where Singapore was once the natural locational choice for Japanese firms. First, all companies expressed some amount of fear that this would affect the situation for their subsidiary or branch office in Singapore per se. Several of the firms note that they have been forced to lay off staff or simply close some of the various company areas, due to the lack of business in Southeast Asia. Secondly, they note the possibilities that might come from a stronger China and from Japanese firms moving there. Hopefully this could lead to increased trade and investments into other parts of Asia and a natural upgrading of the economy in locations that lose competitive advantage in manufacturing. This process enables and strengthens the attempts by the Singaporean government to promote the city state as the natural location for regional head offices in need of managing complex supply chains, secure and well functioning communication network and to be located at a place where the most competitive professional business service solutions are offered (Felker 2003: Yue and Lim 2003). This effort has also been very important for the Japanese companies. JETRO along with interviewed firms all emphasize the suitable environment that exists in Singapore, especially for professional business service firms, due to the level of ICT facilities and the solid institutional environment in the country.

Additionally, the firms benefit from easy living conditions for foreigners, with Japanese schools and the overall usage of English. The well equipped port facilities and Changi airport also strengthen Singapore's prime geographical location, as has the 2002 bilateral free trade agreement between Japan and Singapore. Since the 2004 free trade agreement between the US and Singapore, however, there are concerns in the Japanese business community over the greater advantages gained by US firms.

In the aftermath of the Asian financial crises many of the Japanese companies started to withdraw or reorganize in East and Southeast Asia. According to JETRO Singapore there were two driving forces behind this changing structure. First, the increased number of mergers in the domestic Japanese market due to the financial problems and increasing competition had repercussions on the respective firms' offices in Singapore. Some firms chose to leave the country or regrouped as a new venture. Secondly, the client base and possibilities of continuing profitable business seemed limited, sometimes prompting a withdrawal. In these cases another possibility was to use contacts in the domestic Japanese market and to move some of the information-gathering function to partner firms remaining in Singapore. In this restructuring process the concentration of business operations became an important way of reducing cost. In other words, the number of Japanese employees was reduced and at the same time the number of total employees was reduced, mainly keeping local staff that are less well-paid and easier to lay-off if necessary. In addition to the financial crisis in 1997/98, the prolonged economic difficulties in the Japanese domestic economy have pressured the overseas operations of firms. This had led to price increases in financial services, for example, provided to Japanese companies. According to a manager of one of the leading Japanese banks this created tensions in the beginning, but firms connected to the bank through the *keiretsu* have now come to accept a higher price along with fees for various forms of consulting activities.

Strategic Considerations

Apart from the macroeconomic problems of the region in the late 1990s the Japanese companies expressed concern with other problems of running business in Singapore. These issues might say more about the competitiveness of the Japanese firms, however, than the abilities of the Singapore government to retain its position of being a regional hot-spot. First, it is difficult to attract the best employees. Most Singaporean graduates want to work for the government as admired civil servants. Secondly, if they choose to work for a foreign firm, their first choice is rarely a Japanese company. It is only possible to speculate on the reasons for this, but lower pay and a hierarchical structure with limited career possibilities might be a few explanations. This connects to the impression expressed by a number of the interviewed executives that local employees do not work as hard as employees in Japan and that they do not share the natural loyalty to the company that is also expected in Japan. On the other hand, long working hours do not necessarily mean efficient business operations. The white-collar sector in Japan has been criticized for having low productivity despite the late nights in many Tokyo offices (Porter et al. 2000). According to JETRO Singapore there is a problem with the career path for local employees in many of the Japanese companies. Only one of the major Japanese companies in Singapore is headed by a Singaporean, all others have Japanese managing directors. This is a problem acknowledged also by the Japanese companies in the interviews, but it seems to be difficult to find a solution in the short term. Perhaps the tendency to localize the labour force in the Japanese companies after the financial crisis can help to create better career paths for non-Japanese employees. The favourable geographic location together with the open and liberal economy has been put forward as the most important reason for being in Singapore. Singapore's bilateral trade agreement with Japan has provided incentive for Japanese companies to use Singapore as the regional base outside Tokyo. However, according to the respondents, this has only had a secondary affect on service producing companies. In other words, if trade and investment in the manufacturing sector is helped by a liberal framework this could be translated into additional business for the professional service sector.

Non-tariff barriers have a negative impact on the regional business environment. They hamper the ability of Japanese companies to utilize the regional potential of Singapore to the full extent. This situation is completely different compared with having a regional headquarters in Europe or North America where regional free trade has existed for a long time. However, the respondents in this study are rather pessimistic about the possibility of creating a solid free trade and investment structure in ASEAN and the North Asian countries in the near future, leaving a localized sense of the regional market. Since more of the Japanese companies' clients choose to organize themselves in Asia around a regional headquarters structure, this gives strong incentives for the service providers to follow this organization structure. The companies in the study have different opinions on the implications in terms of regional strategic independence. It seems to be connected to the managerial strategy of the respective main company in Tokyo. For example, several of the subsidiaries explain that Tokyo exert strong influence over the operations in Singapore even if they are seen as rather independent companies based on the organization structure. This might be connected to the fact that Tokyo is located in the region and that the Northeast Asian operations in China and Hong Kong have great impact on the business possibilities in Singapore. The consequence of this structure is that very few Japanese companies offer their complete range of activities in Singapore even if they are a regional headquarters. Instead they try to single out some operations that are viable for the Singaporean market and fit these services into the regional structure to complement services supplied directly to clients through the head office in Tokyo or other important regional offices in Hong Kong or Beijing.

Customers and Competition

All of the respondents in this study acknowledge the importance of Japanese clients for pulling the companies abroad. None of the responding companies have less than 50 per cent Japanese customers. Of these clients, firms related to the service provider through an ownership- or *keiretsu* relationship are often an important share. According to one of the executives in the study, the existent client base will be utilized to the largest possible extent in getting additional business.

The question of expanding the client base was secondary in the reasons to move abroad. However, once in Singapore many of the Japanese firms tried to expand the client base outside their mother company or *keiretsu* group. This process has been accelerated by the Japanese domestic economic problems and the Asian financial crisis. As mentioned above, several companies chose to leave, but the remaining firms have tried to put more energy on attracting new customers. One important market is strong demand from the Singaporean government, especially for high-end technology services such as information technology and telecommunication. Two of the companies in the study emphasized solid and useful contacts with various parts of government. Introducing new services to the existing client base is considered to be the single most important way of sustain and develop client relations. Additionally, it gives a possibility of trying out new services in large client organizations that could in a later stage be promoted to new customers as a viable frame of reference. However, the respondents in the study are concerned about the difficulties of getting local or foreign clients in Singapore, on the bases of having had a long presence in the market, high technology level and knowing the Asian business context well. Apart from using the existing customer base for introducing new products and indirectly expand the market, the respondents put forward the brand name as valuable in expanding business. A well known Japanese brand can help in attracting primarily new Japanese customers. On the other hand, Japanese brand names might be less well-known to local clients or foreign firms in Singapore. Scepticism over using Japanese service providers with a presumed relation to a competitor in the same sector has been negative for Japanese service providers. This is a problem when trying to expand the client base outside Japanese clientele (Ström 2004).

When asked about threats and opportunities for the operations in Singapore, the company executives shared a number of opinions. The number one threat is the growing importance of China. The representatives are worried about Singapore losing out as a major Asian business hub. A northern shift in business focus could severely affect the Japanese firms. If clients move to China, there is simply no point in having large operations in Southeast Asia. The second issue concerns the increased competition among Japanese firms. This is connected to the overall pressure on Japanese companies abroad to show profit. It further implies a great competition from Western service suppliers as clients become more aware of service purchases on the global market. This diminishes the personal contacts and *keiretsu* or group relations. In the end it boils down to a fear of losing the extremely important base of Japanese customers, due to the difficulties of getting non-Japanese customers.

Despite working in intensely competitive conditions, the executives see a number of opportunities. First, China can also bring new possibilities for companies in Singapore if and as Singapore successfully becomes a strong service economy, powered by solid institutional structures and excellent infrastructure. The strong and dynamic overseas Chinese networks in the region could bring advantages, as growing Chinese clients would need assistance in connecting their operations to Japan and the global market. This could generate new opportunities for Japanese service providers with long experience within complex business and logistical networks. Additionally the Japanese companies see an advantage in that they know Asia better than many of the Western firms, and this can help them as the region grows more important in the global economy. Finally, a dynamic Asian economy will require strong and integrated regional headquarters for Japanese service companies.

The mantra that all companies express is that competition is changing and becoming fiercer. Since competition is industry specific it is difficult to generalize, but a number of issues surfaced during the interviews. Japanese companies are the most prominent competitors when it comes to service the Japanese community. Western firms are also considered to be strong competitors, but it is mainly when trying to expand the client base towards non-Japanese clients. Competition increases as it has become vital to assist clients in complex supply chains, financial systems or image building. Relying only on Japanese clients will not be a viable strategy.

Mini-Cases

The aim of this part of the chapter is to visualize the situation for the Japanese professional business service firms in Singapore through discussing three of the studied sub-sectors in more detail. The mini-cases will cover the activities of one leading *sogo shosha*, one of the biggest Japanese advertising agencies, and finally one research institute.

The Sogo Shosha

With a presence in Singapore dating back to the years of WWI, this *sogo shosha* is one of the oldest Japanese companies in Singapore. Its main Singapore activity is the trading function, which over time has assisted domestic business in Japan and with logistical support on the global market. Apart from trading activities, investment operations have become increasingly important. The office acts as the regional headquarters in Asia, but is also responsible for Africa, Middle East, Oceania and India. It provides financial control and dispatches product specialists if needed. However its main focus is daily operations is Singapore, ASEAN and contacts with Japan and Northeast Asia. Forty of the 140-person staff are Japanese. The regional structure is complemented with a centralized strategy for each business area, under the supervision of the regional officer at the Tokyo head office. This creates a dual management structure, in which local markets fit into the regional and the product strategy. Every month, the managing director of the Singapore office travels to Tokyo for a discussion with the head office.

According to the managing director the *keiretsu* connection is important and creates what he calls a 'loose integration'. In the start-up phase of internationalization the *keiretsu* is considered important due to assistance in contacts and procurement. In the case of the *sogo shosha*, it has often acted as support to group companies. The *keiretsu* has a monthly meeting of all managing directors of the respective group companies. In Singapore, 21 of the entire 25 companies within the *keiretsu* are present. About 30 per cent of total worldwide turnover is related to *keiretsu* business. The share is similar for the operations in Singapore. Despite important but mature markets in Europe and North America, Asia represents about half of total overseas profit for the *sogo shosha*. The core business of trading has seen a decrease in profit, because new IT solutions have put pressure on the traditional role of the *sogo shosha* as an intermediary. On the other hand, other activities such as investment in retail have become important for business. Today, 70 per cent of the total customer base is Japanese. The brand name is of great importance for getting Japanese customers.

After the Asian financial crisis there has been a shift in business operations among Japanese firms in Southeast Asia. Manufacturing has been diversified out into the region from Singapore and there was a sharp drop in demand for the services supplied by the *sogo shosha*. The managing director of this Japanese *sogo shosha* sees a number of opportunities. First, the financial strength and capabilities will be of great advantage in assisting clients forming supply chains in the region. This also enables the *sogo shosha* to act as an investor rather than only supplying capital and services. Instead of paying for services the company can get assistance in return for a company share. Secondly, the suitable location and the prominent infrastructure in Singapore enable companies to move more into high-value added services. This could further strengthen the business possibilities for high-end services supplied to firms with an even more evident regional reach. An informal advantage could be the company group representative who is on the board of GIC, the investment company owned by the Singapore government.

In an attempt to create better in-house service efficiency and to provide professional services for external customers, the *sogo shosha* started a new company in October 2003. The role of this company is to give advice to departments on finance, accounting, credit control, and human resources. From the start-up three persons were allocated to the company. This company should not only service the *sogo shosha*, but extend its reach within the *keiretsu* group. Again, this shows the indirect importance and connections of *keiretsu* connections. The future outlook for reaching outside the *keiretsu* in the short run is very limited according to the managing director. At present the head of this company is not the outward person needed in promoting business activities on the open market. The managing director concludes that the company needs to employ people with better social skills in order to really prosper and be head-on competitive.

The Advertising Agency

The advertising agency is one of the biggest in Japan and has a strong position among Japanese clients. Its present organizational structure was set up in 2001. Earlier the company was represented by an agency office in Singapore. In 2001 the parent company in Tokyo decided that a regional office was needed to assist and facilitate business throughout Asia, and created a company with regional responsibility alongside the agency in Singapore. The FY 2003 total turnover for the company in Asia was 43 billion Yen. Taiwan is today the single most important market and the company has great hopes for Vietnam, which has caught the interest of many of the company's main clients. The regional company has 14 employees of which 8 are Japanese expatriates. The local Singapore agency has 65 employees, including only four Japanese expatriates. The agency is not part of any *keiretsu* group, but acts instead independently with customer contracts.

Ninety per cent of the turnover for the company comes from Japan, and between 80-90 per cent of the Singapore turnover is directly related to Japanese customers. The most important sectors for the company are the auto industry, electronics and other IT related products, food and hygiene products. The headquarters in Tokyo plays an important role in business development, since it often has very close contacts with the respective client's headquarters. According to the regional manager this helps to get the regional contracts. Nevertheless, an increased exposure on the global market by clients threatens the business relationship. It is possible that clients will choose an agency with strong capabilities in a rather specific area instead of an overall agency. Since the company's current non-Japanese client base is very small, there is a strong desire to expand. The initial problem is to get the chance to compete for a contract, since potential clients often consider only a few agencies. It has been difficult to become one of the companies of interest to non-Japanese global clients.

Additionally, a rather complicated structure exists between the local agencies, the regional head office and the headquarters in Tokyo. The role of the regional head office is to network the various agencies and assist them in strategic matters for the Asian market and then let the local agencies focus on the daily routine. This could be to create new business possibilities, assist on financial matters and help to coordinate the networks for clients in Asia. According to the managing director the real reason for setting up the regional head office was because many of the Japanese clients were shifting to a regional structure, managed by regional headquarters. Nevertheless, the company is under the authority of the international department at the head office in Tokyo, and it is the Tokyo management that sets the budget targets. Product specialists from the head office in Tokyo often assist in particular projects.

One of the most interesting differences that the company needs to handle outside Japan is that competing accounts cannot be handled within the same agency. In Japan it is common practice that an agency handles competitors' accounts. The fact that the Japanese clients demand to be treated like international clients is a sign of that they have analyzed the competitive situation and will not necessarily work with a Japanese agency. After a severe drop in business activities in the aftermath of the Asian financial crisis, the company now sees interesting opportunities in assisting clients in a regional strategy where China plays an increasingly significant role. Today the Chinese market is almost as large as the rest of Asia, but growing much more rapidly. In terms of competition, the threat from other Japanese advertising firms is very limited. Rather, the so-called mega-agencies threaten with their strong record of creative work, and global brand names. The agency is very confident of using its strong brand name with Japanese clients, but sees a threat of losing contracts when clients are localizing their business. In other words, non-Japanese managers abroad in Japanese companies could mean an increased reluctance to primarily use Japanese agencies. An additional aspect that helps to attract Japanese clients is the all-round capabilities of the agency, in areas such as PR, media and advertising. Western agencies are more specialized.

The Research Institute

Research institutes have a broad range of services. On the one hand they are like think tanks supplying information to clients and to the public, and on the other hand they are consulting companies covering areas such as management consulting and IT related services. This company started its operations in Singapore in 1984 and today has 20 employees. The company is very dependent upon the parent holding company for getting clients or servicing already existing clients. It is primarily client companies working with securities trade and asset management. Over the last ten years, the firm has expanded its services from information gathering and analysis to IT services. Today the largest activities of the Singapore operation are software applications and consulting for the financial sector. However, the office in Hong Kong has an IT support function that extends across the entire Asian business region. The Singapore office lost its management consulting arm in the aftermath of the financial crisis, but no offices were closed in the region.

Today however, China has become more important, and management consulting has increased its market share. The company has been successful in getting deals associated with both Japanese official aid (ODA) investments, but also private investments where Japanese clients want to reorganize the Asian operations. One reason for this success is the rather independent situation in the Japanese business environment and the fact that the research institute company has solid knowledge about Japanese business practice. The single most important customer for the Singapore office is the asset management arm of the group. The firm is not part of any *keiretsu*, but work associated with the parent holding company is very important. Business connections are maintained through joint international location, not only in the same cities but in the same buildings. Additionally, group directors have served as company presidents. There is no real regional headquarters, and the different business areas are under the supervision of the director for international business in Tokyo. The chairman of the Singapore subsidiary is based in Tokyo. Due to the prolonged recession in Japan and the Asian financial crisis, the company lost many of the small and medium sized Japanese financial customers that previously used Singapore as a regional base. The company has now returned into a more stabile business situation with larger clients. All current customers are Japanese, and the company emphasizes serving these existing customers. According to the company management, there is no real basis for attracting western clients other than serving them indirectly through the group's other arms in securities trading or asset management. While the strategy is to provide high quality rather than low price, the company does not have the combination of both hardware and software that many of its main competitors like NEC and Fujitsu possess.

Conclusions and Further Research

A number of conclusions can be drawn from the study using the OLI framework. Firstly, the empirical results show the Japanese firms' relative lack of O-advantages in serving non-Japanese clients in Southeast Asia. Secondly, the L-advantages of Singapore remain strong, despite the challenging competition arising from rapidly growing cities in China, such as Shanghai. Finally, the management relationships related to the I-advantages exhibit a problematic dual management structure in trying to control foreign operations.

The study shows that many of the firms have an unclear ownership specific advantage relative to domestic and other international firms present in Singapore. The findings underline the limited competitiveness experienced by Japanese companies. It naturally has implications for the future, in terms of service industry growth in Japan and Asia. Nevertheless, the Japanese service companies seem to fare much better in Asia than in Europe, because of the extensive activities of Japanese companies in general in the region. The main problem for the Japanese professional business firms is reaching out to new non-Japanese clients, where the brand name is less known. Since brand name and reputations are vital in services (Aharoni 2000), this negatively affects the ownership specific advantage.

In relation to the location specific advantages it is clear that Singapore is still of great importance for the Japanese companies. This is related to the fact that the city-state is a strong financial centre and has managed to attract regional functions of many transnational companies. This is well in line with the government strategy (Yue and Lim 2003). Nevertheless, new competition comes from Northeast Asia through the rise of China. This helps cities such as Hong Kong, Tokyo and Shanghai to attract more foreign companies. In this process, Japanese manufacturing companies have started to relocate labour intensive operations mainly to China. Indirectly, this generates a different structure for the professional business service market, in terms of types of services and geographical spread. The phenomenon of Japanese companies acting as forerunners in regrouping in Asia and taking advantage of changes in the manufacturing sector has also been found in research on the locational choice of Japanese service multinationals (Edgington and Haga 1998). This research furthermore acknowledges Singapore's importance as being one of the main locations in a hierarchical pattern of cities in Asia with a diverse supply of business services (Edgington and Haga 1998; Taylor et al. 2002). The locational effect of service clusters for attracting additional companies and sectors have been underlined in the study. It is primarily not the exceptional supply of skilled labour that has attracted Japanese service companies, but rather the fact that business conditions are good and not 'being there' limits the regional attractiveness and eventual success. These findings contrasts with research giving less meaning to the aspect of 'being there' at a specific location to take advantage of knowledge flows and market possibilities (Power and Lundmark 2004). Again, in the case of the service industry there seems to be important to have a physical presence in order to move with business development.

To tackle the managerial situation and enhancing the internalization advantages, the research has uncovered interesting aspects of the on-going regionalization among Japanese professional business service firms. Based on the interviews, this process has been accelerated by the changing pattern of service clients. Since the Asian financial crisis Singapore has been losing out mainly to China in terms of manufacturing and assembly plants. This has increased the need for more complex supply chain and logistical services along with other forms of professional business services. Due to the good geographical location and solid institutional framework, Singapore has managed to position itself as a service and regional headquarter hub. Many of the Japanese service firms have tried to regroup and work around Singapore as a regional base for Southeast Asia or even the whole of Asia. It became evident in the interviews however, that several firms express a lack of experience in dealing with a rather independent region structure. The fact that Tokyo, where most companies have head offices, is within reach in the region true regional independence is difficult to achieve. This shows similarities with the management structure for Japanese companies in other regions such as in Europe (Ström and Mattsson 2006), and that there are clear differences in how Japanese companies are governed on the global market compared to Western companies.

To answer the question of why the Japanese firms have had difficulties to internationalize successfully, it seems that the traditional Japanese business structure has not favoured service internationalization. The service firms have not had enough strong ownership-advantages to exploit them in a foreign location like Singapore. While the companies that took part in this study were of very different sizes, even the smaller companies had a relation to a larger mother/holding company or a wider *keiretsu*. It is no surprise that service firms often are small, but what is interesting in the case of the Japanese firms is that they all seem to be dependent on a larger organization. A possible explanation might be that it is very difficult for smaller Japanese firms to take a successful step onto the world market. These findings are well in line with the presumption that a different structure for business service supply exists in Asia (Daniels 1998) and that the *keiretsu* connections are changing in character (Hatch 2001; McGuire and Dow 2003). *Keiretsu* relations are limited as measured in real revenue terms, but they are still important as information networks and for an initial client base. Increasingly, all companies related through a mother company structure or *keiretsu* are pressured to make a profit rather than sustaining traditional ties.

The financial crisis of 1997/98 put many firms under pressure since, the investments of small and medium sized Japanese firms in the region declined. This became especially problematic for companies within finance, insurance related services and advertising.

The study clearly shows that a dual management structure seems to appear within Japanese firms trying to work with a regional headquarters. On the one hand the regional office is working with strategy, but on the other hand product specialists and the international department at the Tokyo head office want their say. Working under fierce competition abroad both from Japanese and other professional business service firms is helping to up-grade the competitiveness of the Japanese companies. It is possible that the combination of more regionally independent organizations working together with the most useful networking parts of the *keiretsu* groups (see Kensy, 2001) and a solid knowledge of Asian business, could in the future result in severe competition for the traditional Western service giants.

References

Aharoni, Y. (2000), 'The Role of Reputation in Global Professional Business Services', in Aharoni, Y. and Nachum, L. (eds) *Globalization of Services – Some Implications for Theory and Practice* (London: Routledge).

Amin, A. (1999), '"Placing Globalization," Theory, Culture and Society' (1997), in Bryson, J., Henry, N., Keeble, D. and Martin, R. (eds), *The Economic Geography Reader* (Chichester: John Wiley and Sons Ltd.).

Bathelt, H. and Glückler, J. (2003), 'Toward a Relational Economic Geography', *Journal of Economic Geography* 3, 117–144.

Beyers, W.B. and Lindahl, D.P. (1996), 'Explaining the Demand for Producer Services: Is Cost-Driven Externalization the Major Factor?' *Papers in Regional Science* 75: 3, 351–374.

Bryson, J.R. (2000), 'Spreading the message. Management consultants and the shaping of economic geographies in time and space', in Bryson, J., Daniels, P.W., Henry, N. and Pollard, J. (eds), *Knowledge, Space, Economy* (London: Routledge).

Clark, G.L. (1998), 'Stylized Facts and Close Dialogue: Methodology in Economic Geography', *Annals of the Association of American Geographers* 88: 1, 73–87.

Daniels, P.W. (1998), 'Economic Development and Producer Services Growth: The APEC Experience', *Asia Pacific Viewpoint* 29: 2, 145–159.

Daniels, P.W. (2001), 'Globalization, Producer Services, and the City: Is Asia a Special Case?' in Stern, R.M. (ed.), *Services in the International Economy* (Ann Arbor: The University of Michigan Press).

Dunning, J.H. (1988), 'The Eclectic Paradigm of International Production: A Restatement and Some Possible Extensions', *Journal of International Business Studies* 19 (Spring), 1–31.

Dunning, J.H. (1989), 'Multinational Enterprises and the Growth of Services: Some Conceptual and Theoretical Issues', *The Service Industries Journal* 9: 1, 5–39.

Dunning, J.H. (1998), 'Location and the Multinational Enterprise: A Neglected Factor?' *Journal of International Business Studies* 29, 45–66.

Edgington, D.W. and Haga, H. (1998), 'Japanese Service Sector Multinationals and the Hierarchy of Pacific Rim Cities', *Asia Pacific Viewpoint* 39: 2, 161–178.

Enderwick, P. (1990), 'The International Competitiveness of Japanese Service Industries: A Cause for Concern?' *California Management Review* 32: 4, 22–37.

Erramilli, M.K. and Rao, C.P. (1993), 'Service Firms' International Entry-Mode Choice: A Modified Transaction-Cost Analysis Approach', *Journal of Marketing* 57: 3, 19–38.

Felker, G.B. (2003), 'Southeast Asian Industrialization and the Changing Global Production System', *Third World Quarterly* 24: 2, 255–282.

Gerlach, M.L. (1992), *Alliance Capitalism: The Social Organization of Japanese Business* (Berkeley: University of California Press).

Gertler, M. (2003), 'Tacit Knowledge and The Economic Geography of Context, Or The Undefinable Tacitness of Being (There)', *Journal of Economic Geography* 3, 75–99.

Hatch, M. (2001), 'Regionalizing Relationalism: Japanese Production Networks in Asia', *MIT Japan Program, Working Paper Series, 01.07* (Cambridge MA: MIT).

Hermelin, B. (1997), *Professional Business Services – Conceptual Framework and a Swedish Case Study*, Department of Social and Economic Geography, Uppsala University.

Japanese Chamber of Commerce and Industry (2003), *Membership Directory 02/03*, JCCI Singapore.

Johansson, J.K. (1990), 'Japanese Service Industries and Their Overseas Potential', *The Service Industries Journal* 10: 1, 85–110.

Johansson, J. and Vahlne, J-E. (1977), 'The Internationalization Process of the Firm: a Model of Knowledge Development and Increasing Foreign Commitments', *Journal of International Business Studies* (Spring/Summer), 23–32.

Johansson, J. and Vahlne, J-E. (2003), 'Business Relationship Learning and Commitment in The Internationalization Process', *Journal of International Entrepreneurship*, 1, 83–101.

Keeble, D. and Nachum, L. (2002), 'Why Do Business Service Firms Cluster? Small Consultancies, Clustering and Decentralization in London and Southern England', *Transactions of The Institute of British Geographers* 27: 1, 67–90.

Kensy, R. (2001), *Keiretsu Economy – New Economy?* (New York: Palgrave).

Majkgård, A. and Sharma, D.D. (1998), 'Client-Following and Market-Seeking Strategies in the Internationalization of Service Firms', *Journal of Business-To-Business Marketing* 4: 3, 1–41.

Masuyama, S. and Vandenbrink, D. (2003), *Towards a Knowledge-Based Economy, East Asia's Changing Industrial Geography* (Singapore: ISEAS).

Mcguire, J. and Dow, S. (2003), 'The Persistence and Implications of Japanese Keiretsu Organization', *Journal of International Business Studies* 34, 374–388.

Ministry of Finance (Japan MOF) (2004) *Outward Direct Investment (Country & Region)* Http://Www.Mof.Go.Jp/English/E1c008.Htm.

Nachum, L. (1999), *The Origins of The International Competitiveness of Firms: The Impact of Location and Ownership in Professional Service Industries* (Cheltenham: Edward Elgar).

OECD (2004), *OECD in Figures 2004* (Paris: OECD).

O'Farrell, P., Wood, P.A. and Zheng, J. (1996), 'Internationalization of Business Services: An Interregional Analysis', *Regional Studies*, 32(2): 101–118.

Ono, H. (2001) 'Restructuring Strategy of Japan's Service Sector in The Twenty-First Century', in Masuyama, S., Vandenbrink, D. and Yue, C.S. (eds), *Industrial Restructuring in East Asia, Towards The 21st Century* (Singapore: ISEAS).

Porter, M.E. (2000) 'Locations, Clusters, and Company Strategy', in Clark, G.L., Feldman, M.P. and Gertler, M.S. (eds), *The Oxford Handbook of Economic Geography* (Oxford: Oxford University Press).

Porter, M.E., Takeuchi, H. and Sakakibara, M. (2000), *Can Japan Compete?* (London: Macmillan Press Ltd).

Power, D. and Lundmark, M. (2004), 'Working Through Knowledge Pools: Labour Market Dynamics, The Transference of Knowledge and Ideas, and Industrial Clusters', *Urban Studies* 41: 5/6, 1025–1044.

Sharma, D.D. (1991), *International Operations of Professional Firms* (Lund: Studentlitteratur).

Sjøholt, P. (1993), 'The Dynamics of Services as an Agent of Regional Change and Development', *The Service Industries Journal* 12: 2, 36–59.

Ström, P. (2003), *Internationalization of Japanese Professional Business Service Firms*, Ph.D. Dissertation, No 39/2003, Department of Social Sciences, Roskilde University.

Ström, P. (2004), *The 'Lagged' Internationalization of Japanese Professional Business Service Firms: Experiences From The UK and Singapore*, Göteborg: Department of Human and Economic Geography, School of Economics and Commercial Law, Göteborg University, Series B, No. 107.

Ström, P. (2005), 'The Japanese Service Industry: An International Comparison', *Social Science Japan Journal* 8: 2, 253–266.

Ström, P. and Mattsson, J. (2005), 'Japanese Professional Business Services: a Proposed Analytical Typology', *Asia Pacific Business Review* 11: 1, 49–68.

Ström, P. and Mattsson, J. (2006), 'Internationalization of Japanese Professional Business Service Firms', *The Service Industries Journal* 26: 3, 249–265.

Taylor, P.J., Walker, D.R.F. and Beaverstock, J.V. (2002), 'Firms and Their Global Service Networks', in Sassen, S. (ed.), *Global Networks, Linked Cities* (New York: Routledge).

Wernerheim, M.C. and Sharpe, C.A. (2003), '"High Order" Producer Services in Metropolitan Canada: How Footloose Are They?' *Regional Studies*, 37: 5, 469–490.

Wirtz, J., Lovelock, C. and Islam, A.K.S. (2002), 'Service Economy Asia: Macro Trends and Their Implications', *Nanyang Business Review* 1: 2, 7–18.

Yeung, H.W. (2003), 'Practicing New Economic Geographies: A Methodological Examination', *Annals of The Association of American Geographers* 93: 2, 442–462.

Yeung, H.W. and Lin, G.C.S. (2003), 'Theorizing Economic Geographies of Asia', *Economic Geography* 79: 2, 107–28.

Yue, C.S. and Lim, J.J. (2003), 'Singapore: A Regional Hub in ICT', in Masuyama, S. and Vandenbrink, D. (eds), *Towards a Knowledge-Based Economy, East Asia's Changing Industrial Geography* (Singapore: ISEAS).

PART 3
Knowledge-Based Services and Regional Development

Chapter 9

Towards Post-Industrial Transition and Services Society? Evidence from Turin

Paolo Giaccaria and Vincenzo Demetrio

The evolution of economic systems from industrial to post-industrial has become a dominant theme in the analyses of contemporary economies. The dynamics of sector composition are, in fact, important in the explanation of levels and rates of productivity variation, income and employment of an economy in time. Understanding the tertiarization of an economy is therefore fundamental. However such understanding can be obscured by over-generalization. This chapter develops and illustrates variations in ideal-typical characterizations of post-industrialization.

The Meanings of the Post-Industrial Transition

An analysis of the relevant literature confirms that separate processes, acting on different sectors of the economy, underlie tertiarization processes (Bell 1973; Gershuny and Miles 1983; Castells 1996). These processes interact with varied institutional, social, and cultural backgrounds. Therefore, the effective paths of post-industrial capitalism or tertiarization are extremely diversified (Chiesi 1998; Esping-Andersen 1993; 1999; Regini 2000).

In this chapter we try to interpret structural changes of the Turin economic system within the context of tertiarization models proposed by the literature. In the process, we eschew one-dimensional and deterministic transition models. Logically, interpretative models used to study national states are inappropriate for analysing provincial areas, since the sub-national areas introduce some systematic distortions in the composition of sectors in the national areas that contain them. Nevertheless, it may be a useful exploratory exercise to comprehend the unique aspects of the Turin productive system (TPS) transition process in comparison to national models.

Institutional Orders and Tertiarization Models

We suggest that specific structural mechanisms acting on each sector of the economy are at the base of tertiarization processes. Those mechanisms operate through local or societal institutions, yielding varied processes of tertiarization. Therefore it is incorrect to identify the structural mechanisms with the empirical results of tertiarization (Castells 1996). From the supply side, institutional arrangements influence the freedom of enterprises to restructure, de-localize and manage their own human resources. In countries where labor relationships are strong, firms will have more difficulties in realizing processes of outsourcing, de-location and organizational slimming. In these countries, therefore, industrial employment will tend to be maintained, while tertiary employment will grow with a lower speed in comparison to those countries where labor institutions are less strong.

Espin-Andersen (1990) identified three ideal typical models of institutional orders: liberal, conservative-familiar and social democratic. By considering the interaction between them and the dynamics of tertiary transition, he posits three kind of trajectories of post-industrial development:

1) The *liberal model* typical of Anglo-Saxon countries is characterized by limited intervention of the state in the economy, either through the supply of social services, or through the regulation of the labor market. Therefore enterprises are free to proceed without ties to plants de-location toward countries with lower manpower cost and to wide outsourcing. Therefore, interaction between this kind of order and the pushes toward tertiarization of the economy yields strong de-industrialization, weak employment growth in the business services sector and the creation of a wide market of social and personal services characterized by low salaries in a framework of economic growth and strong social polarization.

2) The *conservative-familiar model* that characterizes the Mediterranean area and continental Europe countries entails a large role for the state in economic regulation and for families in the supply of social and personal services. The interaction between this model and the post-industrial transition may cause a relative maintenance of industrial employment, due to the strong labor market regulation that guarantees workers' positions in the face of enterprises' demands for outsourcing, de-location and rationalization. Tertiary employment, instead, grows at relatively low rates.

3) The *social democratic model* exemplified by the Scandinavian countries, is characterized by strong intervention of the state in the regulation of the labor market and in the supply of social services to families. This intervention yields an intermediate balance between manufacturing and services, with high rates of female participation. Nevertheless, these countries face strong tendencies toward industrial movement to areas with lower labor costs.

By analysing the evolution of employment by sector in the seven more industrialized countries during the 1970s and 1980s, Castells (1996) reached results similar those of Espin-Andersen. Particularly he distinguished two models of tertiarization (i.e. two social-institutional models of the new "informational economy") based on the interactions between post-industrial transition and institutional models. The first one is the *Service Economy Model* of the Anglo-Saxon countries, characterized by: rapid manufacturing decline; more intense expansion of financial services in comparison to other business services related more directly with industry; expansion of social services; general increase of managerial employees. The second is the *Industrial Production Model* proper of Japan and Germany, with less decline of industrial employment than the preceding model. The production services of technicians and consultation show greater increase than financial services, while social services are in general less developed (particularly in Japan).

Countries such as France and Italy occupy an intermediate position. France has maintained an important manufacturing base and has developed social and personal services. Italy has maintained a reliance on automobile manufacturing and has developed networks of small enterprises. On the basis of Espin-Andersen's and Castells's contributions, Ballarino (2000) built a typology of tertiarization models (synthesized in Table 9.1). The author defines the *liberal* and *social democratic* models as post-industrial, and the *corporative-familiar* model (Castells' industrial production model) as neo-industrial. Particularly, the last model refers to the theoretical and empirical elaboration of Harry Greenfield (1966), developed and deepened among the others, by Fuchs (1968), Daniels (1985), Martinelli (1984) and Tosi (1985), which saw in the growth of producer services the main characteristic of productive system transformations.

Regarding Italy, Momigliano and Siniscalco (1980; 1982; 1986) concluded that 'the absolute and relative growth of the employment in the services was due in a large extent to the increase of intermediary services used in the productive system' in particular by 'the industrial sub-system'.

What Kind of Tertiarization for Turin?

In this section we will try to interpret tertiarization phenomena of the Turin economy in the light of the models developed in the literature and synthesized in Table 9.1.

Neo-Industrial Variations

Tables 9.2 and 9.3 introduce employment trends in the five sectors into which the economic activities of Turin Province have been divided, following the classification of Browning and Singelmann (1978). The decades between 1981 and 2001 are the reference period, while the source used is the Census of Industry and Services.

From Table 9.2, we can observe that Turin's employment has moved progressively toward the tertiary sector. The 1990s were characterized by the acceleration of tertiarization of economy phenomena (Table 9.3). The extraordinary growth recorded by the aggregate of producer services during these years is with no doubt tied up to the outsourcing of functions previously help within the industrial enterprises. This choice attempts to create a more flexible value chain and to improve the productivity of human and technological factors. Nevertheless in 2001 the industrial sector occupied over 47 per cent of Turin Province employees. It is therefore improper

Table 9.1 Tertiarization and development (of employment) models of different economic sectors

| | Tertiarization models | | |
	Post-industrial liberal	Neo-industrial corporative-familiar	Post-industrial social democratic models
Sectors Industry	Decline of the traditional industry, with strong de-location and unilateral restructuring.	Stability or light decline, with orchestrated and regulated restructuring.	Decline of the traditional industry, orchestrated and regulated restructuring.
Producer Services	Strong growth, in particular of financial services and of services with low professional level (cleanings, vigilance).	Growth, in particular of technical-engineering services. Privatization of public producer services.	Growth of the services with high scientific contained, based on the high schooling.
Distributive Services	Steady; decline in case of privatizations (public transport). Industrialization of commerce.	Steady, with strong pushes to the privatization (public transport). Stability of small distribution.	Steady. Industrialization of commerce.
Personal Services	Growth of market, characterized by low salary employments. Decline of communities.	Strong role of domestic communities, with scarce state support (division of the job for gender: scarce female share in salaried job).	Strong role of self-servicing, with state support.
Social Services	Growth, in particular health, financed by market and state and organized privately.	Growth, followed by a diminution of public offer (fiscal crisis of the state), development of a private sector with public financing. Important role of the domestic communities.	Strong growth based on the public offering. Important role of the self-servicing.

Source: Ballarino, 2000.

Table 9.2 Employees' trends for sectors of activity

	Employees			
	1981	*1991*	*2001*	Δ (2001–1981)
EXTRACTIVE	1,809	1,672	1,584	-225
Transformative	483,980	390,630	326,022	-157,958
Services:	248,953	310,107	360,443	111,490
Distributive Services	152,661	168,802	167,975	15,314
Producer Services	42,204	72,863	119,469	77,265
Social Services	3,062	12,833	12,284	9,222
Personal Services	51,026	55,609	60,715	9,689
Total	734,742	702,409	688,049	-46,693

Source: Elaborations on Istat data (Census of Industry and Services).

Table 9.3 Trends of percentages and variations of employees for sectors of activity

	Percentages of employees			Employees	
	1981	*1991*	*2001*	Δ relative (2001–1981)	Δ relative (2001–1991)
EXTRACTIVE	0.25	0.24	0.23	-7.57	-12.44
Transformative	65.87	55.61	47.38	-19.29	-32.64
Services :	33.88	44.15	52.39	24.56	44.78
Distributive Services	20.78	24.03	24.41	10.57	10.03
Producer Services	5.74	10.37	17.36	72.64	183.08
Social Services	0.42	1.83	1.79	319.11	301.18
Personal Services	6.94	7.92	8.82	8.98	18.99
Total	100	100	100	-4.40	-6.36

Source: Elaborations on Istat data (Census of Industry and Services).

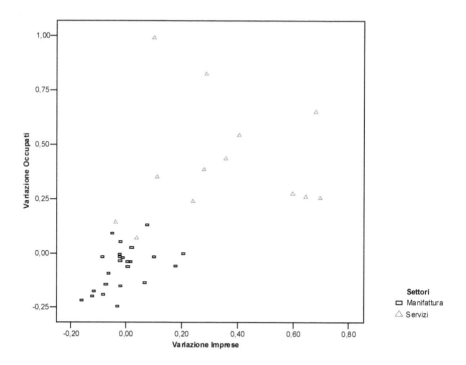

Figure 9.1   Evolution of Turin productive system
Source: Elaborations on Istat data (ASIA).

to talk of "post-industrial" society, since manufacturing continues to represent the engine of the local social-economic system.

We conclude that in Turin the process of tertiarization is characterized by the *neo-industrial model* , with the slight decline of industrial firms and the development of producer services that are connected to industrial rationalization. Nevertheless it differentiates itself from that corporative - familiar proposed by the literature. For example, the growth of distributive services is not tied to the presence of organizational consolidation or industrialization processes. Furthermore, more related to the liberal post-industrial model is the growth of personal and social services.

Evolutionary Typologies of the Local Productive System

We are interested in similarities and differences in the evolution of the local productive system component activities. In fact, even if we thought that the distinction between manufacture and producer service activities is only statistical artifice, nevertheless we believe that the forms of adjustment of the Turin system to the global transformations of capitalism, are different for the two components.

We have developed a data series for manufacturing and service firms' employment 1996–2001. To minimize the effects of possible short period shocks, both for enterprises and employment, three-year averages have been calculated for the periods 1996–1998 and 1999–2001. We then determined the relative variation of each sector (measured as the relationship between the difference of the averages of the two three-year periods). In Figure 9.1 the values of the two variables have been presented in a Cartesian plan reporting on X axis the relative variation of the number of enterprises and on Y axis the relative variation of employment.

Generally, manufacturing sectors were characterized by a general reduction in the number of enterprises and employees, while services showed great dynamism and a general growth in enterprises and employment.[1] Among services, only terrestrial transportation and design and planning show trends comparable to those of manufacturing activities. This fact can be interpreted considering that both transportation and design have been traditionally integrated with the manufacturing system, especially with the automotive cluster which has been dominating Turin production structure since the beginning of the twentieth century.

Analysing individual manufacturing and service sectors, it is possible to identify groups of sectors showing the same behavioral typologies. For manufacturing, (Figure 9.2) four typologies of evolutionary dynamics are identifiable:

1) *Growth/ restructuring* characterized by light employment increases and by small changes (either negative that positive) in the number of enterprises. Relevant sectors of the TPS are pharmaceuticals, plastic objects manufacture, instrumental goods, automotive, publishing.
2) *Multiplication* characterized by an increase of the number of enterprises employment stability. Relevant sectors included aeronautics, mass electronics and shipbuilding.
3) *Steadiness* characterized by stability of the number of enterprises and employment. This describes many manufacturing activities that are important support sectors for the local economy, i.e. base chemistry, siderurgy, metallurgy, apparatuses for telecommunication, rubber articles and other vehicles.
4) *Stagnation* characterized by the diminution of the number of enterprises and employment. This includes many sectors traditionally present in the Turin system as base electronics, precision tools, textile and leather industries, and other manufacturing industries.

---

1   We have excluded the extreme cases of coke manufacture and research and development in social and humanistic sciences. The former was characterized by a huge increase; the latter by a clear diminution of the number of employees.

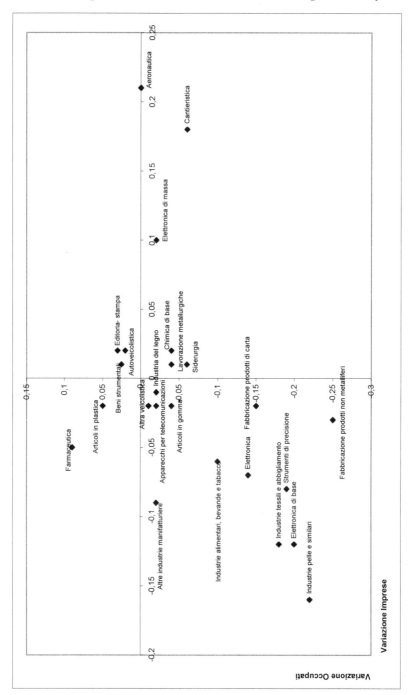

Figure 9.2   Evolution of Turin manufacture system
Source: Elaborations on Istat data (ASIA).

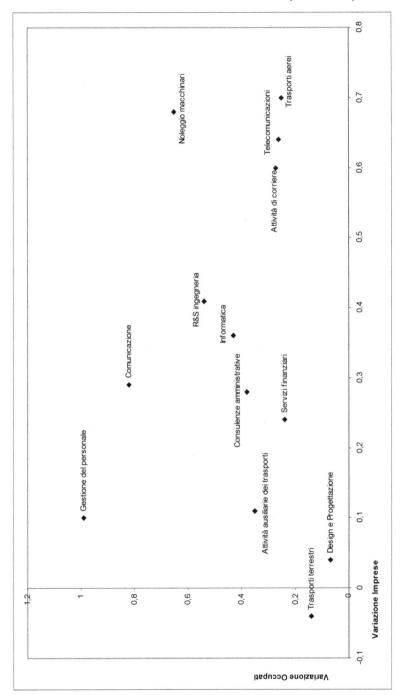

Figure 9.3 Evolution of Turin producer services system
Source: Elaborations on Istat data (ASIA).

The evolutionary dynamics related to service activities (Figure 9.3) are more complex and variegated. Nevertheless some evolutionary typologies could be identified.

1) *Expansion* characterized by high growth rates of employment and by a moderate growth of the number of enterprises, as in the personnel management and communication services.
2) *Development* characterized by growth of the number of enterprises proportional to the growth of employment. This characterizes tertiary activity of great importance for the local economy, such as: research and development in engineering, computer science, administrative consultations, transportation auxiliary activity and financial services;
3) *Steadiness* characterized by stability of the number of enterprises and employment. We found this characteristic in both strategic and support sectors of the local economy such as design and planning and terrestrial transportation.
4) *Multiplication* characterized by large increases of the number of enterprises and by a moderate expansion of employment, as found in telecommunications, delivery services excluding national mails and air transportation.

Empirical Evidence from the Turin Production System (TPS)

In order to evaluate the dynamics between manufacturing activities and related business services a sample of about 500 firms belonging to both sectors was identified as the basis for face to face interviews, based on a structured questionnaire, between April and July 2004. Given the administrative unity of the Province scale (NUTS 4), we have used functionally defined sub-systems, signally the Local Labor Systems (LLS) defined by ISTAT on the base of the commuting home-to-work daily flows.

Table 9.4 shows the geographical distribution of the interviews between LLS. Despite the fact that Turin represents about 60 per cent of the sample – given the still strong feature of geographical concentration of production within the Province – the first comment is necessarily about the uneven distribution of the manufacture/service ratio. If in the overall sample the ratio is about 2:1, in the single LLSs we are facing a clear dichotomy between those systems characterized by an overwhelming manufacturing specialization (mainly in Avigliana and Chieri, but also Rivarolo Canavese, Pinerolo and Cirié) and Turin where almost half of the sample is concentrated in service firms. That is of course not just a sampling trick – or mistake – but a fundamental feature of the territorial economy in Turin Province, where only the main LLS is showing some evidence of "post-industrial" transition. Also it is quite significant that the only alternative pole of concentration of service businesses is the Ivrea LLS – that is the former Italian Silicon Valley, where the long-lasting presence of Olivetti since the 1950s resulted in an ICT district.

Table 9.4 The sample distribution (among Local Labor System in Turin Province)

Sector	Local Labor System (LLS)								Total
	Avigliana	Carmagnola	Chieri	Cirié	Ivrea	Pinerolo	Rivarolo Canavese	Torino	
Manufacturing	28	3	13	21	46	30	23	174	338
	93.3%	75.0%	92.9%	80.8%	63.0%	83.3%	85.2%	59.8%	67.5%
Service	2	1	1	5	27	6	4	117	163
	6.7%	25.0%	7.1%	19.2%	37.0%	16.7%	14.8%	40.2%	32.5%
Total	30	4	14	26	73	36	27	291	501
	100.0%	100.0%	100.0%	100.0%	100.0%	100.0%	100.0%	100.0%	100.0%

Nevertheless, we will focus only on Turin Metropolitan Area (TMA) which is largely overlapping with the respective LLS and we therefore assume the two to be synonymous. We will consider first some general features of the manufacturing-service relationship, to move forward analysing the competitive dynamics of the two sectors and finally considering shortly the relationship between territorial embeddedness and competitiveness.

General Features

Figure 9.4 shows clearly that there is a diachronic gap between the two sectors broadly considered. While manufacturing started its growth (in terms of number of firms) in the 1960s and more evidently in the 1970s to continue in the 1980s and in the 1990s – likely in different specializations rather than those of the 70s – the service sectors' emergence seems to be a more recent fact concentrated mainly in the late 1980s and after 1990. About 80 per cent of the service businesses were established after 1980, against only 50 per cent of the manufacturing enterprises.

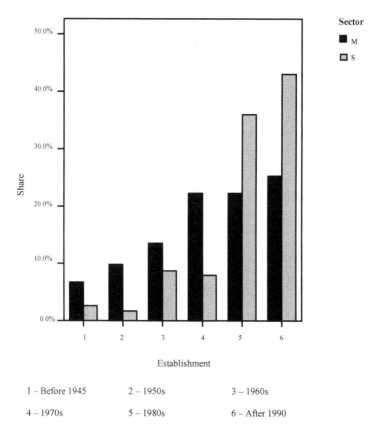

Figure 9.4   **Sample distribution by year of establishment**

Another evidence of the similarity of the TPS with the neo-industrial framework is the variation in enterprises' belonging to a holding or a group. Table 9.5 shows that there is a significant difference between industrial and service enterprises: about 40 per cent of the service firms belong to a group, against 30 per cent of the manufacturing sample. This can be interpreted as an evidence of the process of progressive externalization of services which might have given birth to small units specialized in business-to business services or attracted existing groups to enter the local market. In case of the development of an autonomous service sectors not strictly related to the manufacturing system, in fact, we would expect a higher share of small independent firms, *à la* Silicon Valley.

Table 9.5 Number of firms in the sample belonging to a holding

Sector	Holding belonging		
	No	Yes	Total
Manufacturing	117	53	170
	68.8%	31.2%	100%
Service	68	47	115
	59.1%	40.9%	100%
Total	185	100	285
	64.9%	35.1%	100%

Note: $\alpha < 0.10$

The results about the holding's nationality are even more significant and not ambiguous: in the service sector almost 90 per cent of the groups which are localized in Turin are Italian, while half of manufacturing groups have a foreign ownership (Table 9.6). Once again, this result shows how service sector in Turin has not been linked to the Global Commodity Chain yet and it is therefore likely to be mainly locally oriented. The next section will discuss further this lack of internationalization with regard to the issue on exports.

Table 9.6 Nationality of holding companies

Sector	Holding Nationality		
	Italian	Foreign	Total
Manufacturing	28	28	56
	50.00%	50.00%	100.00%
Service	46	6	52
	88.50%	11.50%	100.00%
Total	74	34	108
	68.50%	31.50%	100.00%

Note: $\alpha < 0.05$

Table 9.7 Perceived competitiveness increase

Sector	Competitiveness Increase		
	No	Yes	Total
Manufacturing	75	98	173
	43.40%	56.60%	100.00%
Service	54	61	115
	47.00%	53.00%	100.00%
Total	129	159	288
	44.80%	55.20%	100.00%

Note: α > 0.10

Competitive Dynamics

The close relationship linking services and manufacturing in the TPS is also evident from the dynamics of competition and competitiveness of the two sectors. Table 9.7 above shows interviewees' perceptions of changes in their firms' competitiveness. We simply asked managers or entrepreneurs if they perceived a shift in their capability to face national and international competition. The results clearly underscore that there is a slight majority with a positive judgment about their competitiveness (55 per cent versus 45 per cent), but there are not significant differences between firms working in service business and those working in manufacturing.

Table 9.8 Negative impact of the Fiat crisis

Sector	Negative Impact		
	Yes	No	Total
Manufacturing	86	86	172
	50.00%	50.00%	100.00%
Service	58	58	116
	50.00%	50.00%	100.00%
Total	144	144	288
	50.00%	50.00%	100.00%

Note: α = 1

This premise of strong relatedness between service and manufacturing in the neo-industrial model can be observed in the Turin case by the perfectly independent distribution of the negative impact on Fiat crisis on sampled firms' businesses (Table 9.8 above). Fiat has been over the past century the main player in the industrialization process in Turin and, to a large extent, in Italy as well. Although Fiat's history has been time by time characterized by recurrent crises since its very foundation, its embeddedness in Turin has been challenged more deeply in the last two decades. Even if this is not the point of our discourse, just a few figures can exemplify why Fiat delocalization has been currently interpreted as the starting of the post-industrial transition in Turin. Fiat employment in Piedmont, for instance, declined from over 100,000 employees in 1971 to less than 30,000 in 2001. Analogously, the number of first tier suppliers was reduced by 80 per cent during the Eighties and the early Nineties (Conti and Giaccaria 2001).

Exactly half of our sample firms have been negatively affected by the crisis of the main Italian car producer and by its parallel location shift away from Turin – two related but distinct features of Fiat restructuring. It is quite interesting to notice the reasons why some firms have been affected or not by the Fiat crisis (Table 9.9). The most surprising issue is that many firms who do *not* complain about negative back-effects from the biggest corporation's predicament are indeed involved in the automobile industry, but have broadened the range of their customers (about one third of the overall sample), or are working exclusively for other producers not located in the TPS (about 15 per cent of the sample). More interesting, only 67 out of a sample of more than 300 firms declare they are completely extraneous to *both* the Fiat supplying system *and* the car industry itself.

In our perspective it is important to notice that there are not significant differences in the answer by service firms – just a slight prevalence of non-car-related businesses – to witness the strong links that are still tying up service sector not only to generic manufacturing, but more precisely to the Fiat supply system.

The strong commitment of the service sector to the overall competitiveness and development patterns of the manufacturing system is also evident if we consider the data on the percentage of exports on overall sales. Table 9.10 shows a clear export tendency among manufacturing firms, with a lower share in service businesses. For instance more than 75 per cent of the service sample do not address their services towards foreign customers at all, and in general fewer than 5 per cent have a ratio of exports bigger than 20 per cent. On the contrary, fewer than 40 per cent of manufacturing firms are entirely oriented towards national markets, while 25 per cent of them export more than 40 per cent (in value) of their products and half of them more than 60 per cent.

Strategic Behavior and Territorial Embeddedness

The first conclusion is therefore that, in the case of the TPS, data confirm a strong relationship between the growing importance of services and the dynamics of the industrial sector, which can be interpreted as a neo-industrial shift rather than of a post-industrial divide. Nevertheless this is only one part of the tale, as it does not determine the territorialization process – that is how firms' competitiveness interacts with geographical process like location and embeddedness. In other terms, from the closeness between service and manufacturing activities it does not necessarily follow that the two sectors share the same degree of relationships within the territory. For instance, let us consider the perceived reasons of firms' increase in competitiveness.

Table 9.9 An assessment of the Fiat crisis back-effect on the TPS

Sector	"I have not been affected by the Fiat crisis as…" [1]					"I have been affected by the Fiat crisis as…" [2]			
	I have no relation with the car industry	I work in the car industry but not within Fiat supply system	My competencies are fundamental for Fiat competitiveness	I do not rely entirely on the Fiat system	Total	I am a first layer supplier	I am a second layer supplier	I work in a related sector	Total
Manufacturing	37	16	5	30	88	38	33	12	83
	42.05%	18.18%	5.68%	34.09%	100%	45.78%	39.76%	14.46%	100%
Service	30	5	3	15	53	26	18	11	55
	56.60%	9.43%	5.66%	28.30%	100%	47.27%	32.73%	20.00%	100%
Total	67	21	8	45	141	64	51	23	138
	47.52%	14.89%	5.67%	31.91%	100%	46.38%	36.96%	16.67%	100%

Note: [1] $\alpha = 0.318$
       [2] $\alpha = 0.586$

Table 9.10 Export oriented manufacturing versus domestic oriented services (value)

Sector	Export share (2003)						Total
	0%	<= 20%	< 40%	< 60%	< 80%	>=80%	
Manufacturing	67	48	13	21	15	9	173
	38.7%	27.7%	7.5%	12.1%	8.7%	5.2%	100%
Service	90	21	0	1	1	3	116
	77.6%	18.1%	0.0%	0.9%	0.9%	2.6%	100%
Total	157	69	13	22	16	12	289
	54.3%	23.9%	4.5%	7.6%	5.5%	4.2%	100%

Table 9.11   Main components of competitiveness increase

		Provincial suppliers	Regional suppliers	National suppliers	International suppliers	Trained workforce	Innovation acquisition	Quality certification	Local policies	National policies	Internationalization	Product innovation	Process innovation	Cooperation	Service externalization
Manufacturing	Mean	3.69	3.46	3.96	3.54	**5.41**	**4.33**	5.25	2.67	2.41	4.24	6.51	6.76	**3.67**	**2.3**
	Median	1.5	1	2.5	1	**6**	**5**	5	1	1	3	8	8	**2**	**1**
	Number	94	94	94	94	**95**	**94**	95	91	92	93	95	96	**93**	**93**
	Std. Deviation	3.10	3.11	3.30	3.13	**3.38**	**3.28**	3.645	2.44	2.28	3.43	3.60	3.35	**3.11**	**2.03**
Service	Mean	3.39	3.07	3.54	3.02	**7.53**	**5.59**	5.05	3.08	3	3.76	6.92	7.29	**5.42**	**3.53**
	Median	3	1	5	1	**8**	**5**	5	1	1	4	8	8	**6**	**2**
	Number	59	59	59	57	**59**	**59**	59	59	59	58	60	59	**59**	**59**
	Std. Deviation	2.62	2.22	2.65	2.46	**2.97**	**3.50**	3.23	2.66	2.54	3.03	3.41	3.36	**3.25**	**2.77**
Total	Mean	3.58	3.31	3.8	3.34	**6.22**	**4.82**	5.18	2.83	2.64	4.05	6.66	6.96	**4.35**	**2.78**
	Median	2	1	4	1	7	**5**	5	1	1	3	8	8	**4.5**	**1**
	Number	153	153	153	151	**154**	**153**	154	150	151	151	155	155	**152**	**152**
	Std. Deviation	2.92	2.80	3.06	2.90	**3.38**	**3.41**	3.48	2.53	2.39	3.28	3.52	3.35	**3.27**	**2.41**

Note: **Bold** = Significant difference between manufacturing and service sector ($\alpha < 0,10$)

Table 9.11 shows the basis for perceived competitiveness increases (one a scale of 1 to 10). Just considering the differences that are statistically significant, we notice that service firms give more importance to hiring a skilled workforce and innovation acquisition from external sources, than do their manufacturing counterparts. It is interesting moreover to note that the overall process of service externalization has been fundamental for service related businesses, but it has not been judged as important by manufacturing enterprises. On the other side, even if not statistically significant, we should notice that – as one could expect – industrial firms rely more on the choice and the quality of suppliers. Nevertheless, the most important result is that the service sector interviews show a strong attitude to consider cooperation as an important source of competitive advantage.

The importance of cooperation in the service sector is largely confirmed by the question about the existence of cooperative linkages with actual or potential suppliers: in this case the share of positive answers among the service firms is equal to 40 per cent versus less than 25 per cent among manufacturers (Table 9.12).

Table 9.12   Presence of cooperative behavior with actual and potential concurrent firms

Sector	Co-operation with concurrent firms		Total
	No	Yes	
Manufacturing	131	42	173
	75.70%	24.30%	100.00%
Service	67	45	112
	59.80%	40.20%	100.00%
Total	198	87	285
	69.50%	30.50%	100.00%

Finally, we asked the interviewed managers and entrepreneurs to evaluate – once again within a range from 1 to 10 – how specifically territorial features influence their competitiveness. Twenty one possible elements have been considered, ranging from traditional location factors like infrastructure and labor market pooling (Table 9.13), to more relational and "cultural" like industrial atmosphere and tradition or effective local development policies.

If we first consider the more traditional location factors, we see that there are some interesting differences:

1) Although the presence of major customers is important for both sectors, it is significantly more important for those operating in the service business. This finding is consistent with what we sustained in the previous paragraph: while the industrial activities are set in an international framework which is now at least partially independent from local customers, the service sector is more strictly tied to the set of local customers. On the contrary manufacturers give

Table 9.13   Assessment of the main territorial sources of competitiveness (traditional location factors)

		Accessibility (infrastructures)	Spaces available for growth	Specialized market labor	Labor force costs	Industrial and union relations	Credit access	Venture capital	Specialized suppliers	Major customers	Local taxation
Manufacturing	Mean	4.52	4.31	4.35	**4.23**	3.43	4.31	3.75	5.31	**5.79**	**3.96**
	Median	5	5	5	**5**	4	5	5	5	**6**	**4**
	Number	172	169	172	**171**	169	168	162	170	**171**	**167**
	Std. Deviation	2.97	2.85	2.85	**2.82**	2.38	2.64	2.44	3.01	**3.23**	**2.94**
Service	Mean	4.17	3.95	4.54	**3.32**	3.44	3.84	3.43	4.74	**6.63**	**3.32**
	Median	5	5	5	**3**	5	5	5	5	**8**	**2**
	Number	115	114	115	**115**	113	115	106	115	**115**	**113**
	Std. Deviation	2.98	2.76	3.17	**2.51**	2.24	2.66	2.26	2.80	**3.53**	**2.76**
Total	Mean	4.38	4.17	4.43	**3.86**	3.44	4.12	3.62	5.08	**6.13**	**3.7**
	Median	5	5	5	4	5	5	5	5	7	3
	Number	287	283	287	286	282	283	268	285	286	280
	Std. Deviation	2.98	2.82	2.98	2.73	2.32	2.65	2.37	2.93	3.37	2.88

Note: **Bold** = Significant difference between manufacturing and service sector ($\alpha < 0.10$)

Table 9.14   Assessment of the main territorial sources of competitiveness (relational and cultural location factors)

		Trust in inter-firms relations	Public Administration interface	Local development policies	Local political continuity	Public research centers	Private research centers	Specialized fairs	Environmental quality of life	Social quality of life	Associations and chambers of commerce	Industrial tradition
Manufacturing	Mean	4.92	3.38	3.76	3.46	3.85	3.29	3.57	3.7	3.66	4.05	5.05
	Median	5	4	5	5	5	3	4	5	5	5	5
	Number	170	165	167	167	167	168	169	166	166	170	169
	Std. Deviation	2.80	2.20	2.49	2.25	2.52	2.23	2.57	2.35	2.40	2.52	3.24
Service	Mean	5.19	3.3	3.31	3.18	4.3	3.79	3.42	3.49	2.96	3.63	4.21
	Median	5	5	3	2.5	5	5	5	5	2	5	5
	Number	114	113	112	114	113	112	113	113	113	113	113
	Std. Deviation	3.48	2.35	2.36	2.19	3.02	2.65	2.47	2.19	2.12	2.44	3.28
Total	Mean	5.03	3.35	3.58	3.34	4.03	3.49	3.51	3.62	3.38	3.88	4.71
	Median	5	4	4	4	5	5	4	5	3	5	5
	Number	284	278	279	281	280	280	282	279	279	283	282
	Std. Deviation	3.09	2.26	2.45	2.22	2.74	2.41	2.53	2.29	2.32	2.49	3.28

Note: **Bold** = Significant difference between manufacturing and service sector ($\alpha < 0.10$)

Underlined = Almost significant difference between manufacturing and service sector ($\alpha < 0.15$)

more importance to the relationship with local suppliers, which is actually one of the main historical feature of TPS.

2) Manufacturers rely more heavily on factors like local taxation (e.g. urbanization costs), the cost of labor and local access to credit,. These factors are less relevant or, in the case of credit, more difficult to access for the service businesses, often SMEs.

3) If we now consider some location factors more directly related to social embeddedness and/or to cultural context we see that, though their importance is lower in comparison to the previous set of factors, some important differences still exist between industrial and service firms. In particular:

   a) Service related enterprises rely more on the presence of public and private research centers than do manufacturers. This is probably linked to the fact that many of them operate in hi-tech sectors, like ICT, design and engineering or research itself.

   b) Manufacturing firms are more sensitive to soft factors (like the social quality of life and the existence of industrial tradition and atmosphere) and policy related factors (like effective local development policies).

Conclusions

As a preliminary conclusion, Turin's transition towards a service economy presents some similarities to the neo-industrial model as defined in the literature. Post-industrial paths are less clear, and the cognitive processes maintain a close relationship with the traditional ones characterizing the former manufacturing specialization. This outcome is largely confirmed by both the quantitative and the qualitative analysis. On the one side, we have noticed how the main growth in employment and plants occurred in production related services. This trend can be better defined considering the important increase in logistics. On the other side, we have witnessed a substantial coincidence in the competitive dynamics characterizing both manufacturing and production services. The two sectors do not differ significantly in terms of ownership or competitiveness. Even more strikingly, the Fiat crisis – i.e. the main outcome of industrial restructuring in Turin and maybe in Italy – has similarly affected both manufacturing and services.

At the same time, our point is that this convergence does not mean that there are not peculiar differences from the neo-industrial paradigm. The statistical analysis has clearly shown that the fastest growing service sectors have been those related to information and communication technology, consistent with the mainstream vulgate that identifies ICT as the engine of post-industrial transition. We also found evidence that service sector growth followed a partially different territorialization path than that which characterized manufacturing. In fact the results of the survey analysis show that there is a stronger commitment of service firms towards establishing cooperative networks within the territory. This is particularly evident in the attitude of service related firms on innovation and research facilities within the metropolitan area. The higher incidence of cooperative behaviors among service firms than among manufacturers (40 per cent versus 24 per cent) can be assumed as another sign of change in the embeddedness process. In summary, this change in cooperative attitude can be likely interpreted as a decrease in the effectiveness of the hierarchy superimposed by the major Fordist company of the area (Fiat), which affected in the past both entrepreneurial attitude and inter-firm relationships in the region.

References

Ballarino, G. (2000), *Il Quadro Generale Del Sistema Dei Servizi Nell'Area Metropolitana Milanese*, Rapporto Di Ricerca, Milano, Ires Lombardia.

Bell, D. (1973), The *Coming Of Post-Industrial Society. A Venture In Social Forecasting* (New York: Basic Books).

Browning, H.C. and Singelmann, J. (1978), 'The Transformation of the US Labour Force: The Interaction Of Industry And Occupation', *Politics And Society* 8, 481–509.

Castells, M. (1996) *The Rise Of The Network Society* (Oxford: Blackwell Publishers).

Chiesi, A.M. (1998), 'La Specificità Della Terziarizzazione in Italia. Un'analisi Delle Differenze Territoriali Della Struttura Occupazionale', *Quaderni Di Sociologia* 42, 41–63.

Conti, S. and Giaccaria, P. (2001), *Local Development And Competitiveness* (Dordrecht: Kluwer).

Dadda, L. (1985), 'Il Livello Tecnologico Dell'area Metropolitana Milanese nel Quadro Tecnologico Internazionale', in *Tecnologie E Sviluppo Urbano* (Milano: Irer, Franco Angeli).

Daniels, P. (1985), *Services Industries: A Geographical Appraisal* (London: Methuen).

Esping-Andersen, G. (1990), *The Three Worlds of Welfare State* (Cambridge: Polity Press).

Esping-Andersen, G. (1993), *Changing Classes: Stratification and Mobility in Post-industrial Societies* (London: Sage).

Esping-Andersen, G. (1999), *Social Foundations of Post-industrial Economies* (Oxford: Oxford University Press).

Fuchs, V. (1968), *The Services Economy* (New York: NBER-Columbia University Press).

Gershuny, J. and Miles, I. (1983), *The New Service Economy: The Transformation of Employment in Industrial Societies* (London: Pinter).

Martinelli, F. (1984), 'Servizi Alla Produzione e Sviluppo Economico Regionale: Il Caso del Mezzogiorno d'Italia', *Rassegna Economica*, N.1.

Momigliano, F.E. and Siniscalco, D. (1980), 'Terziario Totale e Terziario Per Il Sistema Produttivo', *Economia Politica E Industriale*, N. 25.

Momigliano, F.E. and Siniscalco, D. (1982), 'Note in Tema di Terziarizzazione e Industrializzazione', in *Moneta E Credito* N.138.
Momigliano, F.E. and Siniscalco, D. (1986), 'Mutamenti Strutturali del Sistema Produttivo, Integrazione Fra Industria e Settore Terziario', in Pasinetti, L. (A Cura Di) *Mutamenti Strutturali Del Sistema Produttivo, Integrazione Fra Industria E Settore Terziario*, (Bologna: Il Mulino).
Regini, M. (2000), *Modelli di Capitalismo* (Bari-Roma: Laterza).
Tosi, A. (1985), *Terziario, Impresa e Territorio* (Milano: Angeli).

Chapter 10

Post-Secondary Education: Education, Training and Technology Services

Hyungjoo Kim and James W. Harrington

Knowledge has become a primary source of economic growth and development, supplementing or even supplanting replacing natural resources and labor-intensive industries.[1] Knowledge workers, people whose jobs require formal and advanced schooling, are among the fastest growing groups in wealthy countries' workforces (Drucker 1993 and 2002; Nonaka and Takeuchi 1995).[2] Post-secondary education[3] (PSE) has been increasingly emphasized in studies of and policies for economic growth and development.

Universities and colleges, developed from the medieval organizations of cultural conservation, preservation, and knowledge transmission, have added related services to their liberal-arts educational functions. For its part, "industry," defined here as enterprises that produce or supply goods and services for profit, has increased its external links to keep pace with rapid change and complication of the business environment.

These changes have strengthened the link between higher education and economic development. Universities and colleges are recognized as important contributors to national/regional economic development. The cases of MIT and Route 128 and Stanford and Silicon Valley are widely recognized and imitated by many regional economic growth/development policies. While most research on relationships between higher education and industry focus on research-oriented, doctoral-degree-granting universities, this research includes baccalaureate colleges, community

---

1   In the last few decades, we have experienced a transformation into "knowledge economy". It is "knowledge" – not labor, machines, land, or natural resources – that is the key economic asset that drives long-run economic performance. At the heart of this phenomenon lies a complex, multifaceted process of continuous, widespread and far-reaching innovation and technical change (Jaffe and Trajtenberg 2002: 1).

2   Knowledge workers include high-knowledge workers with considerable theoretical knowledge and learning (e.g., doctors, lawyers, teachers, accountants, chemical engineers, etc.) and knowledge technologists whose manual works are based on a substantial amount of theoretical knowledge. The theoretical knowledge necessary for knowledge workers is acquired through formal education, not through apprenticeship (Drucker 2002: 238–239).

3   Formal education that requires or assumes a high-school diploma, through institutions that grant baccalaureate and/or advanced academic degrees.

colleges, and technical colleges. These lower level colleges and universities comprise the majority of PSE providers, and provide important services to their local and national economies.

Post-Secondary Education in the US

On a per capita basis, the US has more universities than most other countries, which have not made as great an investment in their university system (Noll 1998, 10–11). Government support for fundamental research is fragmented and decentralized in the US, emphasizing market-like features of competitive, peer-reviewed research proposals from individual researchers (Noll 1998, 16). In the US, approximately 10 per cent of national R&D effort is conducted in federal agencies and installations, and another 5 per cent in national laboratories managed by private companies, universities, or nonprofit research institutions. About two-thirds of the budget for privately managed national laboratories goes to labs that are run by universities (Noll 1998, 17–18). In addition, private support for universities, whether individual donations or research grants from companies and foundations, is distinctive in the US.

European countries emphasize national laboratories, so that research and education are more separated. In rankings of the leading research institutions within a particular field of science and engineering, most of the world leaders in the US are universities, whereas most of the world leaders elsewhere are government research laboratories. In the US, about 70 per cent of the authors of scientific and technical publications are affiliated with academic institutions. In Europe, university faculty members frequently have association with research laboratories, but these labs are physically separated from universities and the research typically is identified with a laboratory, not a university. In the US, education and research are more integrated, especially in experimental areas of science and engineering that require laboratories. Research and education have strong complementarities. In laboratory sciences, students are crucial to the success of research programs.

Community colleges (so named because their service areas are geographically circumscribed), technical colleges (so named because their focus is on career-oriented technical training culminating in degrees or in non-degree certificates), and junior colleges (so named because they offer few or no degrees beyond the associate level) are important components of the US PSE system. Most, especially those that are government supported (as are most), emphasize three distinct programs:

1) Transfer programs that extend educational opportunities, a cost-effective way to serve the increasing number of students wishing to transfer to baccalaureate or research-oriented institutions (Dougherty 2002, 295; Cohen and Brawer 2003, 21–22).
2) Workforce development programs, a distinctive niche of community and technical colleges. On average, about three-fifths of community college students are pursuing vocational training for immediate entry into the labor

market (Dougherty 2002, 301). Community colleges are actively involved in four dominant fields: health professions, business, protective services (police and fire), and engineering-related technologies. Programs may be variously geared toward: preparing people for their first job; upgrading the skills of current workers or retraining unemployed workers and welfare recipients for new jobs; or providing specific training courses for employers (contract training, as defined by Dougherty (2002, 302; see Cohen and Brawer 2003).

3) Continuing education: a service to interested adults (including those outside the labor force).

Services Provided by PSE Institutions

PSE institutions contribute to industrial development and growth in multiple ways including supplying an educated workforce, occupational training, research collaboration, and spin-offs. While the direct purchaser or client may be individual students, IT-producing industry can also be seen as a direct or indirect client.

Universities and colleges provide training and further education programs for employees of industry. From the industry perspective, rapid technological change and increasing need for diverse skill requirements result in a growing need for external training/retraining of employees as well as internal training. Conventional education alone is not enough for the learning required for the changes, and formal training needs to be complemented by experiential learning[4] (Keane and Allison 1999, 897; OECD 1996).

Given high educational requirements for IT-related jobs and increasing costs for occupational training, universities and colleges are an important provider of occupational training. Firms are willing to interact with universities and colleges to access educational programs (Feller et al. 2002), and various professional degree programs are provided at universities and colleges through building partnerships with industry (Ronalds 1999).

Colleges and Universities as Milieux

At a very general but important level, the faculty, students, and activities of PSE institutions provide an atmosphere of cultural openness and innovation, and a respect for learning, that are attractive to the highly educated professionals required by advanced sectors of the economy (Florida 2002). Recent alumni often prefer to remain in such places, even when other (especially smaller) cities in the same region face serious out-migration of educated people.

---

4   OECD (1996) has identified that a key component of building an effective knowledge economy will be to provide ladders and bridges across the three main areas of learning (academic, educational, and experimental) *Employment and Growth in the Knowledge Based Society*, OECD: Paris.

Labor Supply

PSE institutions provide disciplined courses of study, culminating in academic degrees and expertise that make graduates more productive in professional contexts. Universities in the US are nearly unique in the degree to which they integrate research with education (Innovation in Information Technology 2003, 20–21; Noll 1998, 18–20). An educated labor supply is increasingly important for high technology industries, and recruiting and retaining a pool of educated labor is significant for regional economic growth and development. While the presence of strong PSE institutions does not ensure that graduates will remain in a given region, other graduates are also drawn toward well-educated places. The longer and more specialized the academic training, is the greater the payoff and the necessity for graduates and employers to engage in nationwide or international job searches.

In addition, many academic programs encourage students to participate in off-campus work experience, and many firms and agencies are eager to hire short-term, student interns to augment and refresh their regular workforces. Pursuing an internship while in school is eased by staying in the same local area as one's school.

Occupational Training and Education

PSE institutions provide programs of training and "further education" for current employees. From the industry perspective, rapid technological change and increasing need for diverse skill requirements result in a growing need for external training/retraining of employees as well as internal training. Given high educational requirements for IT-related jobs and increasing costs for occupational training, universities and colleges are an important provider of occupational training. Firms are willing to interact with universities and colleges to access educational programs (Feller et al. 2002), and various professional degree programs are provided at universities and colleges through building partnerships with industry (Ronalds 1999). Smaller employers are much less likely to provide formal worker training (general or specific, on- or off-site[5]), because of tighter operating budgets, working capital, and staffing schedules, as well as generally lower wages that do not encourage long worker tenure (Carnevale and Goldstein 1983; Holtman and Idson 1991; Barron et al. 1997; Harrison and Weiss 1998).

---

5   Using 1987–1992 annual responses from the US National Longitudinal Survey of Youth (thus, getting respondents who were 22–39 over the study period), Veum (1997) found establishment size and firm size to be highly significant determinants of workers' having received on-site, formal training, but not of off-site training.

Collaborative Research

Universities and colleges contribute to basic research awareness and insight for industry partners (Hall et al. 2000). They can focus on long-term research that IT companies are not expected to take to a significant extent (Innovation in Information Technology 2003: 12–14).[6] Corporate R&D is largely focused on product and process development, and it is often difficult for corporations to justify funding long-term, fundamental research whose results are hard to be captured by their investors. "The results of long-term, fundamental research tend to be published openly and thus to become generally known; they tend to have broad value; the most important may be unpredictable in advance; and they become known well ahead of the moment of realization as a product, so that many parties have the opportunity to incorporate the results into their thinking. In contrast, applied research for product and process development can be performed in a way that can be more easily captured by investors: it can be done under wraps, and it can be moved into the marketplace more quickly and predictably" (*Innovation in Information Technology* 2003, 13–14).

Collaborative research is the most widely known mode of interaction between industry and universities/colleges, and technology transfer from universities/ colleges to industry is measured by patent, license, and invention disclosure. Collaborative research includes co-authoring research papers, contract research with individual investigators, consulting by faculty, and certain group arrangements for industry problems, and support and use of research equipment. Universities and colleges also provide sites at which researchers from competing companies can come together to explore technical issues (Innovation in Information Technology 2003, 21).

Entrepreneurial Spin-offs

A growing number of universities and colleges have become involved in and encouraged the incubation of spin-off companies (Florida 1999; Shaping the Future 2001). Spin-off functions of higher education systems have been spotlighted due to their direct contribution to local employment growth and commercialization of academic research (e.g., Steffensen et al. 1999). Successful stories of university spin-offs are widely recognized (e.g., Lycos from Carnegie Mellon University), and relatively more case studies focus on spin-offs, research parks, and business incubators centered on universities and colleges than on other PSE establishments (e.g., Lilischkis 1999).

---

6 "Long-term" research refers to a long time horizon for the research effort and for its impact to be realized. Examples of innovations that required long-term research include speech recognition, packet radio, computer graphics, and internetworking (Innovation in Information Technology 2003: 9).

IT-Producing Industries: Definition and Importance

Information Technology (IT)-producing Industries are defined as industry sectors that supply the goods and services that support IT-enabled business processes, internet, and e-commerce (The Emerging Digital Economy II 1999; Digital Economy 2000 and 2003). IT-producing industries include hardware, software/computer service, communications equipment, and communications services industries. Detailed industry sectors and their SIC and NAICS are listed in Appendix 10.B.

Many information technologies have short life cycles, and producers are intent on quickly getting a product or service to market. Employers often prefer to hire workers skilled in new technologies rather than retrain their current workers (Digital Economy 2000, 45). With IT skill sets closely linked to specific software and hardware technologies, ever-shortening product life cycles create frequent change in the IT skill mix in demand. Specific technical skills often lose value over time, sometimes in as little as two to three years (Education and Training for the Information Technology Workforce 2003, 3). This means that IT workers must acquire new skills frequently in order to maintain their labor market viability and upward mobility. However, given the frequent changes in technology and difficulties in forecasting future skill needs, workers receive little guidance on what training to acquire for long-term success in the IT field (Education and Training for the IT Workforce 2003, 15).

Research Objectives and Design

Given these varied types of services provided by PSE institutions, the ways in which they serve both individual students, governments, and industrial clients, and thus, their roles in regional economic development, it is worthwhile to study the geographical extent and influences of these services. This paper reports on a study of IT-producing firms' use of and interaction with PSE institutions. How distinct are the types, degree, and geography of services provided by different types of PSE institutions? Does geographic distance inhibit different types of interaction differently? Do firms' type and degree of interaction vary by size, sector, or region?

We selected three MSAs based on characteristics of IT-producing industries and universities/colleges located in each area: Seattle, WA; Portland, OR; and Champaign-Urbana, IL (see the list of constituent counties, in Appendix B). According to Markusen's typology (1996) with reference to firm size, interconnection, and local vs. non-local embeddedness, Seattle, WA represents a hub-and-spoke configuration of IT-producing industries – dominated by one or several large and vertically integrated firms surrounded by suppliers. Portland, OR is a satellite platform dominated by large, externally owned and headquartered firms. Champaign-Urbana, IL has relatively small IT-producing industry sectors and is a good example of a university-centered IT complex.

IT-producing companies were selected by SIC as defined in Appendix 10.B. Using an industrial directory, we identified 3,752 establishments in the Seattle MSA, 2,298

Table 10.1   Respondents' size, by MSA

Size	Seattle		Portland		Champaign		Total	
Micro-small (1-9)	18	48.6%	14	48.3%	10	71.4%	42	52.5%
Small (10-99)	11	29.7%	5	17.2%	1	7.1%	17	21.3%
Large (100+)	4	10.8%	10	34.5%	1	7.1%	15	18.8%
N/A	4	10.8%	0	0.0%	2	14.3%	6	7.5%
Total	37	100.0%	29	100.0%	14	100.0%	80	100.0%

Table 10.2   Respondents' sector, by MSA

SIC	Seattle		Portland		Champaign		Total	
Hardware	4	10.8%	8	27.6%	2	14.3%	14	17.5%
Software	29	78.4%	20	69.0%	12	85.7%	61	76.3%
Communication Equipment	1	2.7%	0	0.0%	0	0.0%	1	1.3%
Communication Service	2	5.4%	1	3.4%	0	0.0%	3	3.8%
N/A	1	2.7%	0	0.0%	0	0.0%	1	1.3%
Total	37	100.0%	29	100.0%	14	100.0%	80	100.0%

Table 10.3   Respondents' size, by sector

	Hardware		Software		Communication equipment		Communication service		N/A	Total	
Small (1–99)	7	50.0%	54	88.5%	0	0.0%	1	33.3%	0	62	77.5%
Medium (100–499)	4	28.6%	4	6.6%	1	100.0%	1	33.3%	0	10	12.5%
Large (500+)	3	21.4%	2	3.3%	0	0.0%	0	0.0%	0	5	6.2%
N/A	0	0.0%	1	1.6%	0	0.0%	1	33.3%	1	3	3.8%
Total	14	100.0%	61	100.0%	1	100.0%	3	100.0%	1	80	100.0%

in the Portland MSA, and 248 in the Champaign MSA. In order to compare three case study areas, the same number of IT-producing establishments were selected by stratified selection in each MSA. From three classes by size of establishments[7] and from four groups by SIC (hardware, software, communication equipment, and communication service), 744 firms were selected by the same proportion of each size class and sectoral group within the identified population of each MSA.

A total of eighty IT-producing establishments responded to the on-line survey questionnaire (response rate = 10.8 per cent), between January and May 2004. Tables 10.1–10.3 show the composition of the respondents by MSAs, size of establishments, and SIC groups.

Findings

Technical Employee Recruitment, by Degree Level

First, in terms of recruitment/hiring of technical workers at each degree level, the survey responses indicate that IT-producing establishments are more likely to recruit and hire technical workers from within the same MSAs than from beyond.[8] However, the degree of dependence on graduates of local institutions declines at higher degree levels. Eighty-six per cent hired the largest portion of technical workers with a vocational certificate from institutions in the same MSA, and 97 per cent hired most of their associate-degree holders from the local MSA. Seventy-four per cent of respondents replied primarily on the local MSA for bachelor's degree holders; 65 per cent for master's or professional degree holders; and 52 per cent for doctoral degree holders.

Technical Employee Recruitment, by Establishment Size and Relative Location

We then analysed responses to questions regarding where establishments recruited their technical employees, by size of establishment. The results from the contingency table analyses below show that there exist statistically significant relationships between size of establishments and proximity of their technical-employee recruiting for workers with baccalaureate, master's, or professional degrees. In these cases, large establishments are more likely to interact with universities or colleges outside of the same MSAs where they are located than small establishments. Hiring by small establishments are more likely to be bound within the same MSAs.

---

7   When classifying IT-producing companies by size, the categories were applied to each establishment, not to the entire companies if they were multi-locational operations.

8   The question posed was "Choose a geographic area from which your company has recruited/hired the largest or the second largest portion of technical workers with a degree at each level," with the possible responses being the same metropolitan area, the same state, the same region of the country, within the U.S., or from abroad.

Internship Programs, by Establishment Size

Large establishments are more likely to offer student internship programs than small establishments. Eight out of fifteen (53.3 per cent) respondents with 100 or more than 100 employees offered student internship programs while nine out of forty (22.5 per cent) responded establishments with less than 10 employees and four out of 17 (23.5 per cent) responded establishments with 10–49 employees do. The results of the contingency table analyses show that statistically significant differences of the likelihood to offer student internship programs exist among establishments with different sizes (confidence level = 0.10).

Internship Programs, by Relative Location

The survey results indicate that IT-producing establishments generally hire student interns from universities or colleges located in the same MSAs.[9] However, respondents were more likely to hire interns from local technical or community colleges, and to go somewhat further afield to identify interns from four-year colleges or universities (Table 10.4).

Table 10.4   Regional sources of student interns

	Number of establishments that hired interns from **two-year community or technical colleges** located		Number of establishments that hired interns from **four-year colleges or universities** located	
No interns	32		21	
In the same MSA	9	75.0%	14	60.9%
In the rest of the same state	2	16.7%	5	21.7%
In the rest of the same division	0	0.0%	0	0.0%
In the rest of the US	0	0.0%	2	8.7%
In foreign countries	1	8.3%	1	4.3%
Evenly distributed	0	0.0%	1	4.3%
In other regions	0	0.0%	0	0.0%
N/A	36		36	
Total	80		80	
No. of responses	12	100.0%	23	100.0%

---

9   The relevant questions were worded "Student interns from two-year community or technical colleges are primarily from colleges located in…." and "Student interns from four-year colleges or universities are primarily from colleges or universities located in…"

Internship Programs, by Relative Location and Establishment Size

Geographies of student internship programs are examined by asking the location of universities and colleges from which interns that IT-producing establishments hire came. Large establishments hire a higher proportion of student interns from outside the same MSA where they are located than expected while small establishments are more dependent on four-year universities or colleges within the same MSAs than the rest of the regions as a source of their student interns. The results from contingency table analyses show that a statistically significant relationship exists between the size of establishments and physical proximity of the four-year universities or colleges from which they hired student interns whereas no relationship is found in case of two-year community or technical colleges.

Occupational Training, by MSA

The survey results from IT-producing establishments show that there is a statistically significant relationship among the three MSAs in the likelihood of providing training for their employees (Table 10.5). The respondent IT-producing establishments in Seattle MSA (27 out of 37, or 73.0 per cent) and those in Portland (18 out of 27, or 66.7 per cent) are more likely to provide training for employees while those in Champaign MSA (4 out of 14, or 28.6 per cent) are less likely to provide employee training.

Table 10.5   Responses to the question 'Does your company provide training for employees?'

	Seattle		Portland		Champaign		Total	
Yes	27	73.0%	18	66.7%	4	28.6%	49	62.8%
No	10	27.0%	9	33.3%	10	71.4%	29	37.2%
N/A	0		2		0		2	
Total	37		29		14		80	
No. of responses	37	100.0%	27	100.0%	14	100.0%	78	100.0%

Occupational Training, by Establishment Size

While the responses reported in Table 10.5 could reflect the fact that Champaign has a higher proportion of small respondents (fewer than ten employees) than Seattle or Portland, our comparison of responses does not show any significant increase in the propensity to use training programs (provided by two-year community colleges and by four-year colleges and/or universities) with the increasing size of establishments.

Occupational Training, by Relative Location

Respondents are more likely to use training programs provided by universities or colleges within the same MSA than by those in more distant geographic areas. All sixteen respondents that have used training programs provided by two-year community or technical colleges referred to institutions in the same MSA. Eighteen of the 19 respondents that have used training programs provided by four-year universities or colleges referred to institutions within the same MSA.

Occupational Training, by Establishment Size and Relative Location

Given the overwhelming reliance on local training programs, it is not surprising that the size of establishments does not have a relationship with the locations of universities or colleges whose training programs the establishments use. All of the respondents answered that they used training programs provided by two-year community or technical colleges within the same MSAs where they are located regardless of size of establishments. In case of four-year universities or colleges, all of the respondents except one small establishment used training programs provided by four-year universities or colleges located in the same MSA. The small establishment is a branch operation of a large foreign company, and most of their employees have been trained in the country where the company's headquarters is located.

Collaborative Research, by Establishment Size

A higher proportion of responding establishments with 100 or more employees (six out of 15, or 40.0 per cent) have collaborated with colleges or universities on research projects, than establishments with fewer than ten employees (six out of 41, or 14.6 per cent). The results of the contingency table analysis show that there are statistically significant differences of the likelihood to collaborate with colleges or universities on research projects among small and medium-sized establishments (1–99) and large establishments (100+). We found no relationship between size of establishments and the likelihood to take technical advice from faculty members from colleges or universities.[10]

---

10 The question was worded "Has your company taken technical advice from faculty members (including casual/informal advice and formal consultations) from colleges or universities during the last 12 months?"

A higher proportion of the large establishments responded that they have used equipment and/or facilities of colleges or universities (five out of 15, or 33.3 per cent) than did mid-sized establishments (one out of 17, or 5.9 per cent), or small establishments (two out of 41, or 4.9 per cent). The results of the contingency table analysis below show that there are statistically significant differences in the use of equipment and/or facilities of colleges or universities among large and small establishments.

Concerning technology licensing, it is difficult to find any distinctive trends across different size of establishments because only three out of the eighty respondents answered that they have licensed technologies developed at colleges or universities.

Entrepreneurial Spin-offs, by Region

Overall, 19.0 per cent of the responding establishments have entrepreneurial origins from local colleges or universities (see Table 10.6).[11] Respondents in Champaign MSA were more likely to have entrepreneurial origins from local colleges or universities (six out of 14 establishments, or 42.9 per cent) compared to those in Seattle MSA (seven out of 37, or 18.9 per cent) and in Portland (two out of 29, or 7.1 per cent). Access to UIUC is one reason for IT establishments being located in Champaign, and one of the few locational advantages of IT establishments in Champaign. From interviews, we understand that entrepreneurs of IT establishments in Champaign often have private connections with UIUC (e.g., spouses or family members of UIUC faculty members or employees who tend to locate their workplace in Champaign).

Table 10.6   Respondents whose companies had entrepreneurial origins in local universities

	Seattle		Portland		Champaign		Total	
Yes	7	18.9%	2	7.1%	6	42.9%	15	19.0%
No	30	81.1%	26	92.9%	8	57.1%	64	81.0%
N/A	0	100.0%	1	100.0%	0	100.0%	1	100.0%
Total	37		29		14		80	

---

11 The question was worded "Does your company, specifically this local operation, have entrepreneurial origins in faculty members, graduate students, undergraduate students, or recent alumni from local colleges or universities?"

Conclusions

This research is based on a survey of managers of companies whose major products can be considered information technology, and a perspective that views these firms as "clients" of post-secondary education institutions in their local regions and beyond. From this perspective, these companies make very distinct uses of different types of PSE institutions, in terms of the types of services and their geographic extent.

The services provided by community and technical colleges are more localized than the interns, recruits, training, and collaborations between baccalaureate colleges and especially research universities.

The results from the survey partly support the suggestion that larger IT-producing establishments make more use of the range of services provided by PSE institutions. The tendency for IT-producing establishments to benefit from colleges or universities by means of labor supply and research collaboration increases with the growing size of establishments. Larger establishments hire larger numbers of technical workers trained at colleges or universities and especially have higher degrees of interactions with graduate-level educational organizations (higher proportion of technical workers with a master's or professional and a doctoral degree among their total employees) than small establishments. Larger establishments more frequently collaborate with colleges or universities in research projects and receive technical advice from faculty members at colleges or universities than small establishments. In terms of occupational training, however, the size of establishments does not seem to have any relation with the likelihood for IT-producing establishments to use colleges or universities to train their employees. Finally, entrepreneurial origins from colleges or universities are not related to the current size of establishments.

The smallest of the three metropolitan areas studied, which also had the smallest IT-producing sector, exhibited a lower propensity to work with the local community college and university to provide continuing occupational training. However, the local IT-producing sector has closer entrepreneurial ties to the local research university than in the other two regions studied.

Overall, the research findings support the oft-cited assertions that universities support economic development, and that this support is fairly localized. The findings suggest that lower and non-degree granting institutions (community, junior, community, and four-year colleges) play large roles, as do the doctoral and research universities more often studied in the academic economic development literature. It will be important to survey the PSE institutions in order to understand the nature, impact, and geographic extent of their services to individuals and IT-producing companies.

References

Barron, J.M., Berger, M.C. and Black, D.A. (1997), *On-the-Job Training* (Kalamazoo, Michigan: W.E. Upjohn Institute for Employment Research).

Carnevale, A.P. and Goldstein, H. (1983), *Employee Training: Its Changing Role and An Analysis of New Data* (Washington DC: American Society for Training and Development).

Cohen, A.M. and Brawer, F.B. (2003), *The American Community College* (San Francisco: John Wiley & Sons Inc).

Computer Science and Telecommunications Board (2003), *Innovation in Information Technology* (Washington, DC: National Academies Press).

Dougherty, K.J. (2002), 'The Evolving Role of the Community College: Policy Issues and Research Questions', in *Higher Education: Handbook of Theory and Research Volume XVII*, edited by J.C. Smart and W.G. Tierney (New York: Agathon Press).

Drucker, P.F. (1993), *Concept of the Corporation* (New Brunswick, N.J.: Transaction Publishers).

Drucker, P.F. (2002), *Managing in the Next Society* (New York: Truman Talley Books, St. Martin's Press).

Feller, I. (1999), 'The American University System as a Performer of Basic and Applied Research', in *Industrializing Knowledge: University-Industry Linkages in Japan and to United States*, edited by F. Kodma, L.M. Branscomb and R. Florida (Cambridge, Massachusetts: The MIT Press).

Feller, I., Ailes, C.P. and Roessner, J.D. (2002), 'Impacts of Research Universities on Technological Innovation in Industry: Evidence from Engineering Research Centers', *Research Policy* 31, 457–74.

Florida, R. (1999), 'The Role of the University: Leveraging Talent, Not Technology', *Issues in Science and Technology*, Summer.

Florida, R. (2002), *The Rise of the Creative Class* (New York: Basic Books).

Hall, B.H., Link, A.N. and Scott, J.T. (2000), *Universities as Research Partners*, NBER Working paper 7643 (Cambridge, MA: National Bureau of Economic Research).

Harrison, B. and Weiss, M. (1998), *Workforce Development Networks* (Thousand Oaks, California: Sage Publications).

Henry, D., Cooke, S., Buckley, P., Dumagan, J., Gill, G., Pastore, D. and LaPorte, S. (1999), *The Emerging Digital Economy II* (Washington DC: Economics and Statistics Administration, Office of Policy Development, US Department of Commerce).

Holtmann, A.G. and Idson, T.L. (1991), 'Employer Size and On-the-Job Training', *Southern Economic Journal* 58, 339–355.

Jaffe, A.B. and Trajtenberg, M. (2002), *Patents, Citations and Innovations: a Window on the Knowledge Economy* (Cambridge, MA: Massachusetts Institute of Technology).

Keane, J. and Allison, J. (1999), 'The Intersection of the Learning Region and Local and Regional Economic Development: Analyzing the Role of Higher Education', *Regional Studies* 33: 9, 896–902.

Lilischkis, S. (1999), *Impediments and Promotion of Technology-based Start-ups from the University of Washington and the Ruhr University Bochum*, Ph.D. dissertation, Ruhr University, Bochum, Germany.

Markusen, A. (1996), 'Sticky Places In Slippery Space: A Typology of Industrial Districts', *Economic Geography* 72: 3, 293–313.

National Association of State Universities and Land-Grant Colleges (2001), *Shaping The Future: The Economic Impact Of Public Universities* (Washington, DC: NASULGC Office Of Public Affairs).

Noll, R.G. (ed.) (1998), *Challenges To Research Universities*. Washington D.C.: Brookings Institution Press.

Nonaka, I. and Takeuchi, H. (1995), *The Knowledge-Creating Company: How Japanese Companies Create The Dynamics Of Innovation* (New York: Oxford University Press).

OECD (1996), *The Knowledge-Based Economy* (Paris: OECD).

Roberts, E.B. and Malone, D.E. (1996), 'Policies and Structures for Spinning Off New Companies from Research and Development Organizations', *R&D Management* 26: 1, 17–48.

Ronalds, B.F. (1999), 'Involving Industry in University Education: The Master of Oil and Gas Engineering', *European Journal Of Engineering Education* 24: 4, 395–404.

Steffensen, M., Rogers, E.M. and Speakman, K. (1999), 'Spin-Offs from Research Centers at a Research University', *Journal of Business Venturing* 15, 93–111.

The Carnegie Foundation for the Advancement of Teaching (2001), *The Carnegie Classification Of Institutions Of Higher Education 2000 Edition*.

US Department of Commerce (2000), *Digital Economy 2000* (Washington DC: Economics And Statistics Administration, Office of Policy Development).

US Department of Commerce (2003), *Digital Economy 2003* (Washington DC: Economics And Statistics Administration, Office of Policy Development).

US Department Of Commerce (2003), *Education and Training for the Information Technology Workforce*, Report To Congress From The Secretary Of Commerce, April 2003.

Veum, J.R. (1997), 'Training and Job Mobility among Young Workers in the United States', *Journal Of Population Economics* 10: 2, 219–233.

Appendix 10.A   2003 Combined/Metropolitan Statistical Areas

Seattle-Tacoma-Olympia, WA Combined Statistical Area

Bremerton-Silverdale, WA Metropolitan Statistical Area

Kitsap County, WA

Oak Harbor, WA Micropolitan Statistical Area

Island County, WA

Olympia, WA Metropolitan Statistical Area

Thurston County, WA

Seattle-Tacoma-Bellevue, WA Metropolitan Statistical Area

Seattle-Bellevue-Everett, WA Metropolitan Division

King County, WA

Snohomish County, WA

Tacoma, WA Metropolitan Division

Pierce County, WA

Shelton, WA Micropolitan Statistical Area

Mason County, WA

Portland-Vancouver-Beaverton, OR-WA Metropolitan Statistical Area

Clackamas County, OR

Columbia County, OR

Multnomah County, OR

Washington County, OR

Yamhill County, OR

Clark County, WA

Skamania County, WA

Champaign-Urbana, IL Metropolitan Statistical Area

Champaign County, IL

Ford County, IL

Piatt County, IL

Appendix 10.B   Information Technology-producing Industries

Industry group	Title of industry	SIC	1997 NAICS
Hardware Industries	Computers and computer equipment and calculating and office machines	3571, 3572, 3575, 3577 (part), 3578, 3579 (part)	333311, 333313, 334111, 334112, 334113, 334119, 334418
	Wholesale trade of computers	5045 (part)	421430 (part)
	Retail trade of computers	5734 (part)	443120 (part)
	Electron tubes	3671	334411
	Printed circuit boards	3672	334412
	Semiconductors	3674	334413
	Passive electronic components	3675, 3676, 3677, 3678, 3679 (part), 3661 (part)	334414, 334415, 334416, 334419
	Industrial instruments for measurement	3823	334513
	Instruments for measuring electricity	3825 (part)	334515
	Laboratory analytical instruments	3826	334516
Software/ Computer Service Industries	Computer programming services	7371	541511
	Prepackaged software	7372	5112, 334611
	Wholesale trade of software	5045 (part)	421430 (part)

	Retail trade of software	5734 (part)	443120 (part)
	Computer integrated systems design	7373	541512
	Computer processing, data preparation	7374	5142
	Information retrieval services	7375	51419
	Computer services management	7376	541513
	Computer renting and leasing	7377	532420
	Computer maintenance and repair	7378	811212
	Computer related services, etc.	7379	541519
Communications Equipment Industries	Household audio and video equipment	3651, 3679 (part)	334310
	Telephone and telegraph equipment	3661 (part), 3577 (part), 3679 (part)	334210
	Radio and TV communications equipment	3663, 3679 (part), 3699	334220, 334290
	Magnetic and optical recording media	3695, 3577 (part)	334613
Communication Services Industries	Telephone and telegraph communications	481, 4822, 4899	5133
	Cable and other pay TV services	4841	5132

Source: Digital Economy 2000 and 2003, SIC codes are added from *The Emerging Digital Economy II*, 1999.[12]

12 Two IT-producing industry sectors included in *The Emerging Digital Economy II* (1999), Radio broadcasting and Television broadcasting, were moved into IT-using industry category in the later volume of *Digital Economy* (2002 and 2003) and are not included in this table.

Chapter 11

Danish Regional Growth Strategy in Marginal Areas: Regional Partnership and Initiative

Sang-Chul Park

Research, the creation of knowledge, and the integration of externally generated knowledge are vital to a nation's capacity for innovation. At the same time, it is not just the ability to generate new knowledge that is important for innovation and growth, but also the way in which the knowledge is disseminated among universities, research institutions and businesses, within and beyond the nation-state. This means that a sustainable innovation system at a national level as well as at a regional level plays a most important role in generating economic growth.

The Danish business sector is based on small and medium-sized enterprises (SMEs) that generally have little direct interaction with knowledge institutions such as universities and research institutes. Additionally, these SMEs conduct relatively little research activity due to various internal and external barriers. As a result, it is necessary that the government carry out an innovation policy focused on cooperation among business, university and research institutions.

Denmark, one of the smallest of the OECD member nations with a population of just over five million, has managed well in spite of its few high-tech firms, unfavorable location, few natural resources, and one of the highest production costs in the world (Maskell 2004). Denmark is regarded as a strong knowledge society compared with other nations, especially in Western Europe. Based on strong knowledge capability, the government focuses on generating economic and employment growth to enhance a high standard of living and to maintain the welfare state. Denmark's economy focuses on niche markets for high value added products in all industrial sectors instead of high-tech capability in certain strategic industrial areas. This strategy is widely regarded as a wise policy decision for a small nation trading with medium and large sized nations (Dalum and Villumsen 1994; Dalum, Laursen and Villumsen 1998). To maintain Denmark's standing as one of the least regionally disparate nations in the OECD, the government prioritizes policies to disseminate the economic growth from the center areas to the marginal areas as much as possible.

This chapter focuses on how a knowledge-based economy affects balanced regional economic growth and vitality, and illustrates how the Danish government has structured regional policy to strengthen the knowledge-based economy in marginal regions.

Theoretical Debates

Maskell (2004) has suggested that the business community in a small nation like Denmark has very strong elements of a village, in that opportunistic behavior is likely to be noticed. In Denmark, actors in all sectors know each other either directly or indirectly. Even in the plastic production and furniture manufacturing industries with their large number of small and medium-sized firms, Kautonen (1996) found that producers have a remarkable degree of knowledge of other domestic producers, their main domestic and foreign suppliers and the most important customers. Maskell et al. (1998) argued that sectoral clustering in Denmark reflects the advantage of proximity not only in supply cost or in low lead-time, but also in learning.

Clusters or industrial districts are not designed but developed in certain areas in which local rivalry stimulates entrepreneurial spirit and reinforces productivity. Saxenian (1994) explained that firms trust each other if most believe that opportunism will be punished. In sectoral clusters, Maskell and Malmberg (1999) suggested that proximity could create a village atmosphere where malfeasance is punished and at the same time, trust relations can be formed and used for sustained knowledge creation. Maskell (2001) observed that competition between firms in clusters stimulates entrepreneurial spirit and reinforces productivity. Firms may compete while helping each other overcome technical problems in the regions and nations where the majority believe that opportunism is penalized.

With increasing international competition, the demand for knowledge exchange and new network relations between the firms seem to be higher than ever before. Therefore, it is significant to create a trust-enhancing environment in order to develop new products and processes. The trust-creating environment makes co-operation possible for the firms, which have little barriers for interaction and exchange of knowledge. Maskell (1999) pointed out that these kinds of informal institutions play important roles particularly in the knowledge-based economy.

Along with a learning culture and trust-creating environment, inter-organizational competence contributes to sustained economic development. This includes widely observed routines and conventions for identifying possible business partners, coordinating, and understanding steps taken in business relationships. Aydalot (1986) argued that shared history, values and culture ease certain types of exchange and operation. Dei Ottati (2002) suggested that the inter-organizational competence is enhanced by the shared culture of the nation or region such as the village atmosphere based on the proximity in the cluster.

This chapter adopts an eclectic theory focused on learning culture, trust, cluster building, and inter-organizational competence to explain that under what circumstances Danish regional growth strategy can be effectively operationalized.

Regional Growth Strategy in Denmark

Background

The Danish government aims to create strong economic and employment growth to achieve a high standard of living and to maintain the welfare state. The main economic challenge facing the government is to simultaneously raise productivity and generate high employment. To carry out overall growth strategies, the government focuses on five economic aspects where special attention is needed at the national level. These are companies' access to labor, investment and capital provision, dynamics and entrepreneurship, research and development, and free and open markets (Ministry of Economic and Business Affairs 2002a). These targeted areas are equally significant. However, Denmark is still in a weak position particularly in the field of research and development compared to other areas due to its small and medium-sized firm industrial structure.

Useful knowledge can diffuse or be purchased across national borders. As a result, a small nation like Denmark benefits from the international creation of knowledge. However, the best way to receive new knowledge is through internal R&D, which improves the capability to acquire knowledge from other nations (OECD 2000). R&D and innovation affect productivity. However, these processes are long and accompanied by many failures although even a few successes result in a high level of profit.

The Danish central government targets a reasonable economic balance between all regions, which contributes to economic growth. The government aims to develop all 14 regions to be attractive areas in which to live and do business although the economic balance between the regions is already regarded as high. The regional growth strategy is carried out by improving the general conditions for growth that include lower tax rates on employment income, efficient capital and labor markets with fewer administrative burdens, and favorable conditions for education and research (Ministry of Economic and Business Affairs 2003).

Overview of the Regional Growth Strategy

Denmark is divided into 14 regions and 275 local authorities. Among these, the cities of Copenhagen and Frederiksberg are unitary authorities: at once regions and local authorities. By international standards Denmark, along with other Nordic nations such as Norway and Sweden, has achieved a high degree of regional economic balance. As a result, the income disparity between the cities and hinterlands in Denmark shows less than in other European nations (see Figure 11.1 and Table 11.1). However, at the same time, marginal areas still exist. Fifteen minor towns and their hinterlands have relatively low earned incomes, populated by nine per cent of the total population. Therefore, the government makes efforts to promote better conditions for doing business and to exploit settlement opportunities in order to generate regional economic growth in these marginal areas.

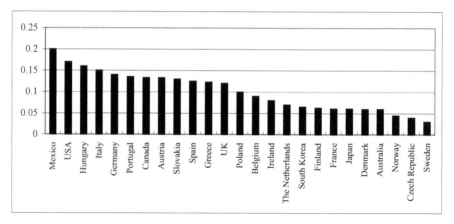

Figure 11.1   Regional disparities in GDP per capita in 25 OECD nations
Source: OECD's Territorial Development Policy Committee, Geographic concentration and territorial disparity in OECD countries 2002.

Table 11.1   GDP per capita in the cities relative to surrounding areas (1995–2000)

Metropolis / Hinterland	1995	1996	1997	1998	1999	2000
Metropolitan area / the rest of Denmark	**1.38**	**1.36**	**1.36**	**1.34**	**1.34**	**1.36**
Glasgow-Edinburgh / the rest of Scotland	1.31	1.34	1.35	1.38	1.39	1.39
Berlin / the rest of North-East Germany	1.59	1.50	1.45	1.43	1.39	1.39
Stockholm / the rest of Central and Southern Sweden	1.31	1.33	1.39	1.42	1.44	1.45
Helsinki / the rest of Southern Finland	1.42	1.45	1.41	1.48	1.50	1.51
Dublin / the rest of Ireland	1.48	1.50	1.50	1.54	1.54	1.54
Amsterdam / the rest of the Netherlands	1.53	1.54	1.57	1.60	1.64	1.62
Hamburg / the rest of North-West Germany	1.80	1.83	1.85	1.87	1.90	1.92
Munch / the rest of Bavaria	1.94	1.97	1.98	2.01	2.04	2.00
Brussels / the rest of Belgium	2.23	2.28	2.26	2.29	2.28	2.29

Source: Ministry of Ec. and Business Affairs: Danish Regional Growth Strategy 2003.

Among 275 local authorities, 15 areas are least developed based on economic disparity. These areas are located in seven of the 14 regions. Funen County and North Jutland County each has four areas; Storström County and Århus County each has two areas. The other three areas are located in Bornholm County, South Jutland County, and Viborg County.

In 2003, the government launched the Danish Regional Growth Strategy, joining local and regional governments and businesses in targeted initiatives to strengthen economic growth by enhancing the conditions for innovation, entrepreneurship and human resources. The key goals of the regional growth strategy are to spread economic growth of the regional centers to the marginal areas, and to create a framework for sustainable economic growth in the outer regions based on regional competitiveness. To generate favorable conditions for regional growth and balance, the government focuses on the following factors: lower tax rates in the marginal areas, smooth and efficient capital and labor markets, fewer administrative burdens, and favorable conditions for education and research (Ministry of Economic and Business Affairs 2003). To support regional competitiveness, the Ministry of Economic and Business Affairs, the Ministry of Food, Agriculture and Fisheries, and the Ministry of the Interior and Health have dedicated budgets of 60 million DDK, 25 million DDK, and 16.9 million DDK respectively (Ministry of Economic and Business Affairs 2003).The targeted initiatives are regional growth partnerships, increased transport allowances in outer regions, and changes in equalization transfers.

These initiatives have generated several important changes. Firstly, nine regional partnerships among the central and regional governments have been established in the outer regions. Each growth partnership draws up a development strategy and launches initiatives to solve problems and difficulties in the region. The government provided financial support to regional initiatives in innovation and digital infrastructure in 2004. Secondly, the transport allowance for long-distance commuters living in marginal areas has been raised since 2004. Lastly, 500 million DKK (about 83.3 million US dollars) will be provided to disadvantaged municipalities for equalization transfers from 2004 (Ministry of Economic and Business Affairs 2004).

A small nation like Denmark is limited in developing high-tech initiatives across all sectors as large advanced nations have attempted. However, a small nation can focus on a few specific strategies perhaps based on niche markets. For Denmark, these strategies consisted of applying information technologies to traditional industries.

The central government's strategic goal is to encourage knowledge-oriented and value-added technologies in all industrial sectors instead of high-tech capability in certain strategic industrial areas. Eight broad industrial sectors (see Table 11.2) are targeted. These Danish sectors have an above-average market share that contributes to developing the national economy and strengthening national competitiveness. These eight sectors comprise about 90 per cent of all economic activity in Denmark. These strongholds account for 93 per cent of value added, 87 per cent of total export, and 89 per cent of employment in the national economy. The resource areas consist of a value-chain of interrelated industries: upstream suppliers, supporting industries, and main customers (Drejer et al. 1999). By supplying a favorable environment of demanding and supporting entities, the competitiveness of the cluster of firms should increase. These sectors are the growth centers at the national level, and the national government has taken steps to disseminate the benefits of economic and industrial growth to the marginal areas at the regional level.

Table 11.2   Identified industrial resource areas

Industrial area	Value added (%)	Employment (%)	Export (%)
Food	14	14	21
Construction, housing	13	15	7
Medico/health	3	2	4
Transport/ communication	11	12	20
Environment/ energy	6	4	6
Tourism/ leisure	6	6	3
Consumer goods	4	6	3
Service	33	30	23

Source: P. Maskell, 'Learning in the Village Economy in Denmark', 2004.

Analysis of the Regional Growth Strategy

Statistical Overview

In order to analyze the regional growth strategy it is useful to examine the average personal income in the 15 areas. The personal income in the areas is a good indicator of whether the balanced regional growth strategy has been implemented properly or not (see Table 11.3). Furthermore, it is also significant to analyze productivity, employment level, labor force participation rate, and demography in the areas in order to understand differences of the areas.

The personal income in the 15 areas is less than the national average, although the income growth rate in many of these areas is higher than the national average. In 2001 the average income growth in the 15 marginal areas was 4.47 per cent while the national average growth was 4.2 per cent. In 2002 the average income growth rate in the 15 areas was 5.33 per cent, while the national average growth rate was to 4.5 per cent. Additionally, the number of areas that achieved lower income growth rate than the national average decreased from five in 2001 to three in 2002 (see Table 11.4). This indicates that the regional growth strategy was implemented properly to a certain extent although the personal income in the marginal areas was not yet met the national average.

Productivity, employment levels, labor force participation rates, and demography in the 15 areas vary greatly, influencing the regional income level. Overall however, the marginal areas are characterized by low productivity, an aging population, low labor force participation rates among working-age people and high rates of unemployment (see Table 11.5). Due to the different challenges faced by the marginal areas, it is reasonable and prudent for the government to have carried out the regional growth strategy based on regional initiatives and competitiveness.

In most marginal areas the economic structure is dominated by the primary sector, craft, retailing, transport, tourism and the public sector. Small companies predominate in traditional industries that are characterized by older technologies.

Table 11.3 Average personal income in the marginal areas (2000–2002) (1,000DKK)

Area	2000	2001	Growth Rate (%)	2002	Growth Rate (%)
Nakskov	164.1	172.9	5.4	184.9	6.9
Nykøbing. F.	181.6	188.5	3.8	198.7	5.4
Rønne	180.9	187.7	3.8	196.8	4.8
Marstal	169.4	178.5	5.4	188.5	5.6
Rudkøbing	171.5	178.7	4.2	190.1	6.4
Svendborg	183.6	191.3	4.2	203.3	6.3
Ærøskøbing	171.4	180.9	5.5	189.6	4.8
Tønder	194.4	199.2	2.5	207.5	4.2
Grenaa	189.6	197.2	4.0	206.2	4.6
Samsø	177.0	188.5	6.5	193.1	2.4
Morsø	184.9	194.8	5.4	200.1	2.7
Fredrikshavn	184.5	192.2	4.2	203.3	5.8
Hjørring	189.2	198.0	4.7	210.4	6.3
Læsø	176.5	181.0	2.5	193.0	6.6
Skagen	187.7	197.0	5.0	211.2	7.2
National average	**203.8**	**212.3**	**4.2**	**221.8**	**4.5**

Source: Danmarks statistik, *Statistical Yearbook 2002, 2003, 2004.*

Table 11.4   Average income growth rate in the nation and marginal areas (2000–2002)

	Marginal areas	National average	Lower income growth areas than national average
Average growth rate in 2001 (%)	4.47	4.2	Nykøbing F., Rønne, Tønder, Grenaa, Læso
Average growth rate in 2002 (%)	5.33	4.5	Tønder, Samsø, Morsø

Source: Author's own adaptation based on *Statistical Yearbook 2002, 2003, 2004.*

Table 11.5   Decomposition of four factors in the 15 areas (2001)

Areas	Productivity	Employment level	Participation rate	Demography
Skagen	-7.0	-3.8	-0.1	-0.6
Hjørring	-6.5	-2.4	0.6	-3.7
Morsø	-7.3	0.6	0.7	-6.1
Tønder	-8.1	0.3	0.7	-5.1
Grenå	-4.7	-2.6	-1.8	-3.8
Frederikshavn	-8.1	-4.4	-1.2	-2.0
Svendborg	-7.9	-3.0	-4.6	-1.6
Nykøbing-Falster	-5.7	-2.5	-8.0	-3.1
Samsø	-8.4	-4.1	-1.1	-8.6
Rønne	-9.9	-4.5	-4.9	-4.1
Ærøskøbing	-8.4	-1.9	-3.2	-11.3
Læsø	-10.1	-8.9	-2.2	-6.6
Nakskov	-7.6	-4.5	-14.4	-4.7
Marstal	-11.8	-2.9	-9.0	-7.4
Rudkøbing	-12.3	-5.2	-6.5	-7.8

Source: Ministry of Economic and Business Affairs, the Danish Regional Growth Strategy 2003.

Notes:
Productivity = primary income per employed person
Employment level = employment to labor force ratio = 1 – unemployment rate
Participation rate = the ratio of labor force to population between 16–66 years of age
Demography = % deviation from the national average in the ratio of population between 16–66 years of age to total population

Usually, technology services, and research partnerships are not prevalent in these low-tech firms. However, Samsø, one of the marginal areas, encourages technology adoption and industrial restructuring by launching regional initiatives for product innovation within existing firms. With IT technologies implemented by a regional partnership the community has been converted from a small rural area based mainly on fisheries and tourism to an attractive island focused on a growing IT industrial sector as well as a tourism industry providing all information in its website (Ministry of Economic and Business Affairs 2002b).

Success Stories

The successful story in Samsø is based on a range of exciting IT projects as well as business and community cooperation. Samsø was the first marginal area in Denmark to obtain broadband internet. In addition, Samsø farmers' association is the first organization to make use of community software developed for Samsø IT. This enables farmers to download news, receive warnings about crop problems, offer items for sale and send emails. From the business side, the Samsø business forum has established IKT Samsø ApS, which has created a call center for telephone answering service and telemarketing. Furthermore, the Association of IT Samsø has been created on the island to promote public information about IT and counts 170 members who are individual companies and households. As community initiatives, the daily Samsø Post operates a new portal for the tourist office and cooperates with the Samsø ecological museum for providing GPS-controlled tourist information (Ministry of Economic and Business Affairs 2003).

The extension of education programs in Bornholm is another example of local initiative. Because of Bornholm's special regional situation near Swedish territory, the Danish Ministry of Education has allocated a special fund for educational initiatives. Presently, seven education programs have been extended to Bornholm. These are a teacher education program, pre-school teacher education program, social worker program, needlework teacher program and three short-cycle education programs. So far, student enrollments have been limited. However, the Glass and Ceramics School in Rønne on Bornholm is the successful local initiative supported by the Ministry of Culture. The school was founded in 1997 and has provided the three year arts and crafts program each year. Unlike other educational institutions on Bornholm, the Glass and Ceramics School has been able to attract students not only from Denmark but also from other Scandinavian countries (Ministry of Economic and Business Affairs 2003).

Conclusions

Overall, the eclectic theory, with its emphasis on learning culture, trust, cluster building and inter-organizational competence, is useful in understanding Danish regional development strategy and its effects. At the same time, however, the result of economic activities in 15 marginal areas illustrates that the similar backgrounds do not guarantee uniformly positive results. The difference in economic performance among these 15 areas is quite large.

Denmark does not hold any global high-tech industries and corporations. For a small nation, Denmark has achieved tremendous development in terms of industry and income per capita, which has contributed to creating a balanced regional growth. Instead of high-tech industries, the nation has developed niche-market oriented industrial sectors in eight resource-based areas such as food, housing, transport, services etc. Individual SMEs are reluctant to carry out R & D activities owing to various internal and external reasons. Therefore, the promotion of interfirm initiatives and linkages is an important task for the government in order to create a sustainable regional development for the marginal areas in particular.

Denmark, along with other Nordic nations such as Norway and Sweden, has among the least regional disparity of the OECD member-nations. Despite such low regional disparity, Denmark has 15 marginal areas exhibiting weak industrial activity and creating low income per capita, which are behind the national average. Therefore, the main target of the Danish regional growth strategy focuses on developing the 15 areas by allocating resources such as education, soft infrastructure, and subsidy for commuting. The key point of the regional growth strategy is based on an efficient dissemination of the economic growth from the center areas to the marginal areas as much as possible. Additionally, the marginal areas are keen to restructure and upgrade their existing industries for the sustainable development by using information technologies. This has been implemented through regional partnerships and financial supports to regional initiatives in innovation and digital infrastructure to strengthen the knowledge-based economy in the margial areas. The Glass and Ceramics School on Bornholm and IT projects on Samsø are representative cases of indigenous development strategies.

The success of these measures is perhaps indicated by the average income growth in the marginal areas having been higher than the national average in the recent years. This may also reflect the fact that the former is still behind the latter so far. The regional development strategy must be reinforced continuously. This means that the central government has to promote the marginal areas strongly, even as they develop themselves based on regional initiatives and strategies.

In sum, Denmark seems to be succeeding in creating balanced regional development based on a strategy to develop niche technologies across a wide variety of sectors.

References

Aydalot, P. (1986), *Milieux Innovateurs en Europe* (Paris: GREMI).

Danmarks statistik (2002), *Statistical Yearbook 2000* (Copenhagen: Danmarks statistik).

Danmarks statistik (2003), *Statistical Yearbook 2001* (Copenhagen: Danmarks statistik).

Danmarks statistik (2004), *Statistical Yearbook 2002* (Copenhagen: Danmarks statistik).

Dalum, B. and Villumsen, G. (1994), *National Industrial Development and Competitiveness* (Copenhagen: Erhvervsfremme Styrelsen).

Dalum, B., Laursen, K. and Villumsen, G. (1998), 'Structural Change in OECD Export Specialisation and Stickness', *International Journal of Applied Economics* 13: 3, 423–443.

Dei Ottati, G. (2002), 'Social Concentration and Local Development: The Case of Industrial Districts', *European Planning Studies* 10: 4, 449–466.

Drejer, I., Kristensen, F.S. and Laursen, K. (1999), 'Studies of Clusters As a Basis for Industrial and Technology Policy in the Danish Economy', in OECD (ed.), *Boosting Innovation: The Cluster Approach* (Paris: OECD), 293–314.

Kautonen, M. (1996), 'Emerging Innovative Networks and Milieux: The Case of the Furniture Industry in the Lahti Region of Finland', *European Planning Studies* 4: 4, 439–456.

Maskell, P. (1999), 'Globalization and Industrial Competitiveness: the Process and Consequences of Ubiquitification', in E. J. Malecki and P. Oinas (eds), *Making Connections: Technological Learning and Regional Economic Change* (Aldershot: Ashgate), 35–60.

Maskell, P. (2001), 'Towards a Knowledge-based Theory of the Geographical Cluster', *Industrial and Corporate Change* 10: 4, 919–941.

Maskell, P. (2004), 'Learning in the Village Economy of Denmark', in Cook, P., Heidenreich, M. and Braczyk, H-J. (eds), *Regional Innovation Systems* (London: Routledge), 154–187.

Maskell, P. and Malmberg, A. (1999), 'Localised Learning and Industrial Competitiveness', *Cambridge Journal of Economics* 23: 2, 167–186.

Maskell, P., Eskelinen, H., Hannibalsson, I., Malmberg, A. and Vatne, E. (1998), *Competitiveness, Localised Learning and Regional Development: Specialization and Prosperity in Small Open Economies* (London: Routledge).

Ministry of Economic and Business Affairs (2002a), *Conditions for Growth in Denmark* (Copenhagen: MOEBA).

Ministry of Economic and Business Affairs (2002b), *Progress Report on Samso IT Lift* (Copenhagen: MOEBA).

Ministry of Economic and Business Affairs (2003), *The Danish Regional Growth Strategy* (Copenhagen: MOEBA).

Ministry of Economic and Business Affairs (2004), *Structural Reforms in the Danish Product and Capital Markets* (Copenhagen: MOEBA).

OECD (2000), *STI Outlook 2000: Denmark Case* (Paris: OECD).

Saxenian, A. (1994), *Regional Advantage* (Cambridge, MA: Harvard University Press).

Chapter 12

Urban Revival and Knowledge-Intensive Services: The Case of the English 'Core Cities'

Peter Wood

Context

The geography of the 'New Economy' is commonly thought to have been expressed mainly through the emergence of 'new industrial districts', or their various synonyms (Scott 1988; Harrison 1992; Markusen 1996). As a result, researchers and policy-makers have become preoccupied with fostering clusters of interdependent activity, hoping to reproduce the innovative success over the past 30–50 years of exemplary regions such as Silicon Valley. The Porter 'branding' of the cluster concept has reinforced this process, while incorporating a wider array of conditions, (Porter 1998; 2000; Martin and Sunley 2003). 'Evolutionary' concepts have also been adopted to explain the cumulative development and transformation of regional economies. These focus on collective learning between key actors and institutions supporting technological change, arising from processes of historical continuity, social networking and chance (Boggs and Rantisi 2003; Bathelt and Glückler 2003; Bathelt and Boggs 2003). On the other hand, concern has been expressed to avoid the 'lock-in' consequences of clustering that have caused the downfall of many industrial regions in the past. Such cases illustrate the importance for regional adaptability of the quality of external, as well as internal production and market relations. Successful regions, whatever their indigenous potential, must also adopt imported technologies and attract inward, 'branch plant' investment. Outside production and market intelligence is an essential component of knowledge exchange. The most innovative intelligence is likely to be personalized and tacit, but much becomes more codified as competitive production evolves (Sturgeon 2003). These 'extra-regional' sources of regional competitiveness are, of course, becoming more significant with economic globalization.

Complementarity between the 'local' and the global' is especially important for urban regions as they come increasingly to depend on knowledge-intensive activities (Amin and Thrift 1992, Swyngedouw 1996, Scott 1998; 2001; Storper 1997; Yeung 1998). If there has been a 'new economy' in Europe over the past 20–30 years, it is concentrated into such regions, and their dynamism has depended

Table 12.1   GDP per capita and regional innovation scores, 2001

	GDP pc 000 Euros	Compared with national 000 Euros p.c.	Regional Innovation scores	Rank (/25)
Frankfurt am Main	74.5	49.4	115	9
Paris	*67.2*	*42.4*	160	3
Munich	61.4	36.3	150	4
Stuttgart	53.6	28.5	145	6
Brussels	*51.1*	*25.1*	111	15
Amsterdam	38.2	10.9	110	16
Stockholm	*35.7*	*10.7*	225	1
Helsinki	35.3	9.3	215	2
London	**35.1**	**7.6**	**150**	**4**
Edinburgh	35.1	7.7	102	20
Milan	32.1	10.1	115	9
Glasgow	31.9	4.5	102	20
Bristol	**29.4**	**2.0**	**140**	**7**
Lyon	29.0	4.2	115	9
Dortmund	26.5	1.4	90	22
Rotterdam	26.2	-1.1	115	9
Leeds	**25.6**	**-1.8**	**82**	**24**
Turin	25.0	3.0	115	9
Toulouse	24.9	0.1	135	8
Birmingham	**22.1**	**-5.3**	**115**	**9**
Manchester	**22.1**	**-5.3**	**104**	**19**
Newcastle	**20.5**	**-6.9**	**85**	**23**
Lille	20.2	-4.6	70	25
Barcelona	18.4	0.4	105	18
Liverpool	**16.5**	**-10.9**	**110**	**16**

GDP estimates for core city local authority areas, 2001.
Innovation scores for administrative regions centred on core cities, 2002.

on such activities. UK developments have been particularly concentrated into London and nearby regions, but continental growth has been more dispersed across the urban system. This situation provides the context for the recent 'Core Cities' initiative in the UK, which is attempting to address the comparative lack of modern economic success in the major English cities outside London compared with similar non-capital cities elsewhere in Europe. The initiative has focused on a varied group: Manchester, Birmingham, Leeds, Liverpool, Sheffield, Newcastle, Nottingham and Bristol.[1] These were the core English manufacturing/port cities of the industrial revolution, and also of post-World War II recovery. Like London, however, they have been badly affected by deindustrialization since the 1970s.

Recent comparative data suggest that non-capital continental cities are now generally more prosperous and productive, and also lead their national economies in these respects (Office of the Deputy Prime Minister 2004). In contrast, the English cities appear to have lower per capita GDP than those in Germany, Scandinavia and Italy, and many in France and the Netherlands, and also lag behind national productivity. Table 12.1, from the ODPM Report, compares the data for some English core cities (excluding Sheffield and Nottingham) with a range of European comparators.[2] In general, capitals, and the larger trading and manufacturing cities, especially in Germany, performed best, but most of the English core cities have less than one third their GDPs per capita. They compare poorly even with the Scottish cities, Edinburgh and Glasgow, and are also consistent in falling behind national productivity levels, unlike most of their continental counterparts.

Other measures, of workforce skill levels, connectivity, and attractiveness to business, suggest possible reasons for this productivity disparity. On the evidence of Table 12.1 the persistent legacy of manufacturing dependence and decline seems to underlie these. The core city regions also appear not to have attracted or developed significant new-technology sectors. The EU's regional 'Innovation Scoreboard' is based on education levels, employment in high technology manufacturing and services, and R&D and patent activity in the regions including the cities (European Union 2002). From these measures, only the Bristol region has a technological profile that might resemble the most innovative continental cities.[3] Some regions around continental cities with comparably low per capita GDPs, such as Toulouse, Turin, Rotterdam and Lyon, at least show better innovative potential. These data may be revealing, but the fortunes of cities are more complex than any one-dimensional explanation might suggest. Technological innovation, especially, can only ever be one part of urban or regional economic success, and may not even be essential to it.

In broader 'evolutionary' terms, the core cities need to foster a new suite of tradable activities around which to focus their key actors, institutions and relationships. It seems unlikely that these will be based in any general technological or manufacturing revival. In the UK, these functions fled the cities forty years ago, primarily to the competing attractions of areas such as the 'M4 Corridor', west of London, or around Cambridge. Facing modern economic restructuring, core cities must promote other knowledge-based assets to mould their future national and global

---

1 The core city study does not include Scotland, with its separate administrative arrangements.

2 In Table 12.1, the city data are based on a private client survey by Barclays Bank, and national data from Eurostat.

3 The Index for Greater London itself is 100.

role.[4] This chapter is based on a preliminary analysis of the scale and structure of these assets. The core cities' economic prospects need to be built on service-based, mainly non-technological, adaptability and innovation. New technologies must be embraced, but will rarely come from distinctively local sources. London seems to have made the necessary transition, through its unique national and international position. The core city studies argue that the other English cities must seek to make the most of their potential. Their future will probably require a more conscious coordination of various private and public sector initiatives, for example to ensure an adaptable, trained, high income, socially inclusiveness workforce, or to improve transport connectivity.

Service Skills and the Adaptability of Cities

Regional productivity and competitive advantages reflect general knowledge-based capacities to respond to change in all its forms (Beyers 2002; Wood 2002b). A diverse range of expertise is required across various local agencies and institutions to select, adapt and apply necessary skills to serve complex market outcomes. Technological skills are only one component, even for technologically-initiated innovation.[5]

London's economic revival over the past 20 years has been based essentially on a reassertion of its historic range and flexibility of commercial and public sector expertise, responding to growing international market opportunities. This has been lead mainly by business and financial services, transportation and trade, and the cultural and consumer services, including tourism. Success has depended on an institutional environment that encourages international investment, attracts and effectively deploys a large, well trained and experienced labor force, and encourages competitive management and marketing practices. Technological expertise has been important, but overwhelmingly directed to adapting international information and communications technologies (ICT) to market requirements. London's size and diversity also mean that its competitive environment can support a great deal of failure, and trial and error learning.

---

4   The UK Office of the Deputy Prime Minister (ODPM) has recently identified six 'drivers' of urban competitiveness: innovation, human capital, economic diversity, connectivity, strategic decision making capacity, and quality of life factors (ODPM 2004). These all relate to the capacity to sustain and ability to exploit a wide range of knowledge-based functions, combining local assets effectively with outside ideas, capital and expertise.

5   Information and communications technology (ICT) activities themselves in the UK are dominated by services such as computer-related activities, telecommunications, wholesaling and renting. In 2000 these contributed a GVA at current basic prices of £46bn, compared with less than £14bn from ICT manufacturing. About half of ICT value added goes to intermediate demand in other sectors, a quarter is exported, and one seventh invested, mainly in the service sector. Although much smaller than ICT goods trade, ICT services also showed a £2.8bn export surplus (1.8 exports /4.6 imports), compared with over a £9bn goods deficit (48.0/38.7) (UK Input-Output Analyses 2002, Office of National Statistics, Section 1).

The English core cities, like their continental equivalents, lack London's scale, scope and degree of slack.[6] They thus need to make the most of those assets they do possess by combining the expertise of large and small firms, service and manufacturing sectors, and private and public agencies. In achieving this coordination, continental cities may have benefited from their relative administrative and financial autonomy (Le Galès 2001; Simmie et al. 2004). Another advantage may be the high quality of public investment in physical, welfare and cultural infrastructure, shown in many business surveys to be important in attracting and retaining urban investment. The need for effective urban economic governance is probably even greater in England, where the national urban system is so dominated by the capital. During the latter half of the twentieth century, however, English cities became more dependent on central government finance and development directives than in any comparable country.[7]

More generally, since the 1980s, UK governments have placed more emphasis than most EU member states on neo-liberal restructuring. This may have brought macro-economic benefits, in low inflation and measured unemployment, and more labor flexibility, but the prevailing geographical outcome has been greater regional income inequality. The Greater South East (GSE) has been favored, including London, the South Eastern region and adjacent areas of East Anglia, the Midlands and the South West. Employment here has not primarily benefited because it includes the most innovative manufacturing sub-regions of the UK. Neither has any international technological leadership been created, since the UK remains a relative laggard in the commercial development of innovation (Department of Trade and Industry 2003). Instead, GSE growth demonstrates the geographical consequences of a deregulated, politically centralized knowledge economy, boosting the nexus of service exchanges in and around London and other regional hotspots. This has increased their economic dominance and control over the rest of the UK, and extended their international, including global reach.

Since Allen suggested in the early 1990s that a new service-based 'regionalized mode of production' was increasingly detaching the South East from the rest of UK, the trends he identified have continued, through knowledge-based financial and business services, high technology manufacturing and services, consumer-orientated production, and public sector investment in defense, R&D and infrastructure (Allen 1992; Simmie et al. 2002). This is the challenge faced in reestablishing the status of the English core cities. What is their current capacity to respond?

---

6   The populations of the defined core cities in 2001 ranged from 976k and 819k each for Birmingham and Manchester to 381k and 439 in Bristol and Liverpool. Central London had 1.4 millions (see Annex 2).

7   UK cities during the 1980s became, 'the most severely constrained in the whole European Union' by financial centralization and privatization (Le Galès 2001, 250). For example, although there were similar shifts in other countries, including Germany, by 1995 UK local taxes raise only 14% of non-borrowing income, compared with over 40% in Scandinavia, 54% in France, and 25% in Italy.

The Knowledge-Intensity of the Core City Regions

Analysis of the knowledge-based adaptability of regions is obviously fraught with problems and, in practice, comparisons of the knowledge base of the core cities and their regions can be made only by using employment data as surrogates.[8] It is not yet possible to use even these measures for detailed international comparisons within Europe.[9] Such data were available at the time of writing from the Annual Business Inquiry (ABI) using 4-digit sector-based SIC classes for 1998-2002.[10] The following components of the 'knowledge economy' in the core city regions were defined:

1) knowledge-intensive business services (KIBS)
2) 'creative' activities, including publishing, computer and other technical services, artistic and media activities and advertising
3) high technology manufacturing and services (based on Butchard 1987)
4) higher education.

The detailed categories are listed in Appendix 12.A, with the core city and city region areas, central London and Greater London, defined in Appendix 12.B. Some computer services and architectural/ engineering services were included in both the KIBS and creative categories, reflecting their double significance for the modern knowledge economy. The aggregated data in the figures for 'All knowledge-intensive activities' include employment in these sectors only once, to avoid double counting. ABI data also allow a comparison of shifts between genders and full- and part-time work, but the analysis here is confined to total employment (see Wood 2002a, Chapter 7). Employment measures could also be improved by occupational data for each sector, showing the actual profiles of locally employed expertise, but these are not available at this scale.

Such measures of the 'knowledge economy' are far from satisfactory. All economic sectors require expert and knowledgeable personnel and it could

---

8   The study for which an earlier version of this analysis was a part was supported by the UK Office of the Deputy Prime Minister. The project for Manchester City Council and the London Assembly was lead by Professor James Simmie, with Professor John Glasson, of Oxford Brookes University, and incorporates economic analysis by Experian Economic Consultants, as well as inputs from Professor Ron Martin, University of Cambridge.

9   Eurostat is developing an Urban Audit across the EU, including those countries that joined in 2004, to enable comparability of data between 258 cities with populations of more than 50,000. So far (late 2005) only general and incomplete data for resident economic activity are available, with no workplace data (Eurostat Press Release, 25 June 2004).

10   The Annual Business Inquiry employee analysis, based on a firm sample survey, has been undertaken on the same basis since 1998. Earlier data are available, having been 'rescaled' to be comparable with post-1998 data, but are less reliable for small-scale analyses.

be argued that the expertise available within established manufacturing firms, and in some circumstances even the consumer and public services, may be of greater significance for local economic success than the presence of specialist knowledge-intensive activities. More localized enquiries would certainly be needed to examine, for example, whether law firms, included in KIBS, actually make a greater contribution to urban business quality and competitiveness than engineering and automobile plants not included in the 'high technology' component. Nor is everyone employed in the defined activities particularly knowledgeable. The significance of their presence, however, it that their core functions should be primarily directed to creating 'new' knowledge.

This is, of course, generally accepted for technologically innovative manufacturing. If it is true of the knowledge intensive business, creative or intellectual services, however, it has a triple significance for local economies. First, if successful, being expert labor-intensive, their own employment should expand. They also influence change among local clients, possibly enhancing their competitiveness and chances of success. Further, if services are sufficiently specialized and innovative, like innovative manufacturing they may develop markets elsewhere, even abroad. They can thus support export-based regional growth, as the key financial and business services do in London. These are the various benefits now sought through the promotion of the business and creative services, and higher education, in the core cities.

Core Cities and London Knowledge-Intensive Employment 1998–2002

In employment terms, the data presented in Figure 12.1 demonstrate the gap between London and the core cities. On average, all the measured 'knowledge economy' activities employed 14.6% of the workforce of the core city regions (CCR: cities and regions combined) in 2002, only half the proportion in Greater London (28.3% in central London and suburbs combined: Figure 12.1a). KIBS are especially dominant in central London (Figure 12.1b), making up three-quarters of its knowledge-intensive activities, compared with two thirds in the core cities, and London has a similarly higher share of the creative sectors (Figure 12.1c). In contrast, the core cities, and especially their regions, had relatively more working in high technology activities, especially around Nottingham, Bristol and Manchester (Figure 12.1d). The share of higher education employment in some of the CCRs, especially Nottingham, Bristol and Manchester, is also greater than London, although not in Sheffield, Leeds or Birmingham (Figure 12.1e). Among the CCRs, Bristol has the largest share of each of the knowledge functions (19.9%), except higher education. Otherwise the Manchester (16.1%) and Leeds (15.4%) city regions are more knowledge-intensive than the others, while Sheffield (9.7%) is least so.

Little long-term significance can be given to the detailed pattern of change between 1998 and 2002 (Figure 12.2). Not only is it too short for any trends to

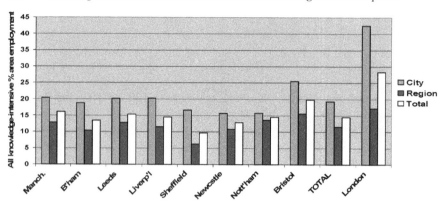

Figure 12.1a   **All knowledge-intensive sectors**

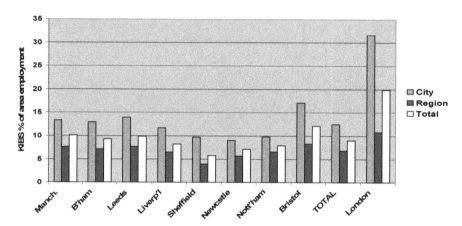

Figure 12.1b   **Knowledge-intensive business services**

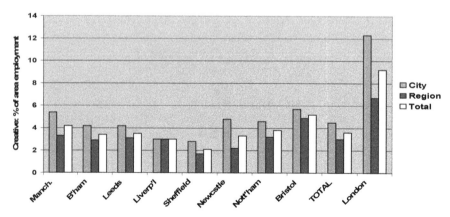

Figure 12.1c   **Creative sectors**

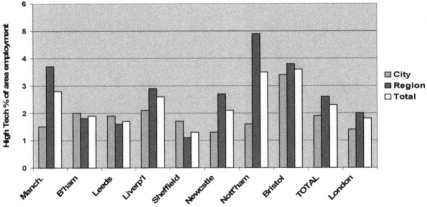

Figure 12.1d   High technology sectors

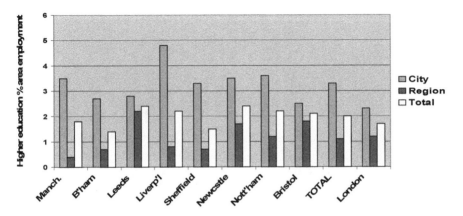

Figure 12.1e   Higher education

Figure 12.1   Core cities, city regions and London: Knowledge-intensive sectors: employment shares, 2002

have been set, but the period also crosses an important cyclical downturn, after the peak of the 'dot-com boom' in 2000, to a trough in 2002–2003. This especially affected London's financial and business services, and might have been expected to be mirrored in the other cities. Year-to-year employment trends are also likely to be affected by single corporate decisions affecting individual office locations, especially in the smaller core cities. Nevertheless, the urban employment response to this cycle indicates the structural contrasts among the cities, as well as net counterurbanization from the cities to surrounding regions.

Figure 12.2 shows that, over the four year period, total CCR knowledge-intensive activities grew at about the same rate as Greater London, by 7–8% (Figure 12.2a).

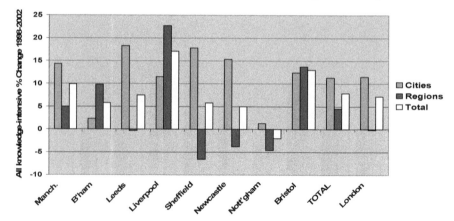

Figure 12.2a   **All knowledge-intensive sectors**

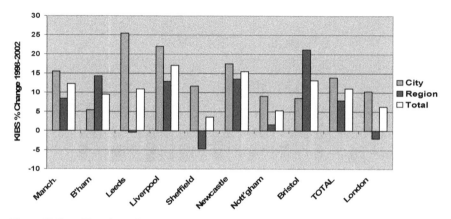

Figure 12.2b   **Knowledge-intensive business services (KIBS)**

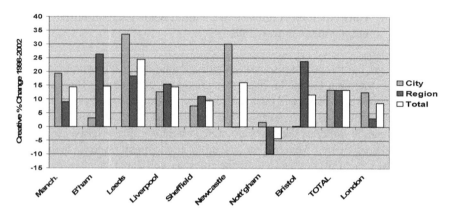

Figure 12.2c   **Creative sectors**

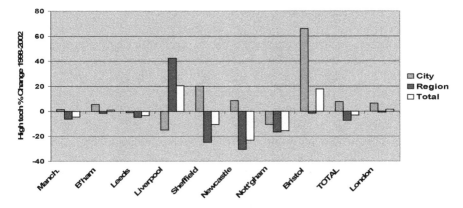

Figure 12.2d   High technology sectors

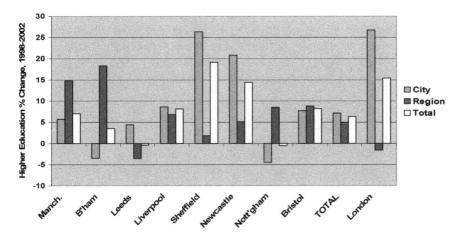

Figure 12.2e   Higher education

Figure 12.2   Core cities, city regions and London: Knowledge-intensive activities: % change, 1998–2002

Net expansion was concentrated mainly into the core cities themselves, however, on average at a similar rate to central London (over 11%). There was particularly high growth in and around Liverpool and Bristol, and in the cities, but not the regions of Leeds, Sheffield, Newcastle and Manchester. In contrast, the two Midlands cities, Birmingham and Nottingham, hardly grew at all, although the Birmingham region offered some compensation. The Nottingham hinterland, like that of Sheffield and Newcastle, actually lost significant proportions of jobs. Compared to central London, core city growth was greatest in KIBS (13.9% cf. 10.3%), but hardly enough to dent the capital's established dominance (Figure 12.2b), while London's higher education employment actually expanded more than any core city in this period (Figure 12.2e).

The different employment responses to the cyclical change around 2000 is shown in the year-to-year data for core city KIBS alone in Figure 12.3. KIBS are here defined at the slightly broader 3 digit level than in Figure 12.1 and 12.2. The differences between London and aggregate core city trends appear to be small (3a), but after 2000 they nevertheless resulted in 23,000 fewer London jobs by 2002, especially in banking, computer services and advertising, and 25,000 more in the core cities. The increase most affected the two largest northern cities, **Leeds** and **Manchester** (12.3b). Both gained over 11,000 KIBS jobs during the four years, in spite of an apparent setback in Leeds in 2000–2001. Manchester's growth was more broadly-based than Leeds', which particularly expanded the financial services and data processing. **Birmingham's** slow KIBS growth was due to sharp losses in 1998–1999, affecting insurance, business services and technical testing, and again in 2001–2002. **Bristol**, which had been a focus for KIBS growth in the previous 20 years, also showed a flat trend during this period. Among the smaller centres, **Liverpool** benefited most from KIBS growth, adding over 5,000 jobs, mainly in insurance and business services. **Newcastle** and **Sheffield** suffered setbacks in 2001–2002, but still gained significantly over the four years. **Nottingham**, after KIBS losses up to 2000, made some recovery, but showed little overall growth by 2002.

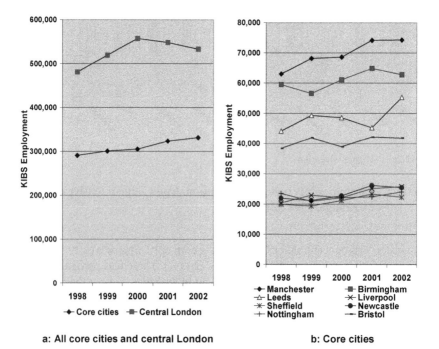

a: All core cities and central London   b: Core cities

Figure 12.3   KIBS annual employment trends, 1998–2002; a: All core cities and central London; b: Core cities

These data show an upward trajectory of KIBS employment in all the core cities, in spite of some local setbacks, with an average trend after 2000 that defied the recession affecting London and London-dominated national employment. More recent data suggest that this continued for at least the next two years (Wood 2006). The growth, however, was particularly focused into the northern cities, Liverpool, Leeds, Manchester and Newcastle, suggesting that the dynamics of their KIBS growth are different from the capital's. In contrast, KIBS employment trends in the more southerly cities, Birmingham, Bristol and Nottingham, more reflected London's fortunes, although showing perhaps surprisingly sluggish growth even before 2000. They thus appear to play a different role from the northern cities in the national disposition of London dominated KIBS activities. More evidence for these patterns comes from examining each group of core city knowledge intensive functions in more detailed.

Knowledge-Intensive Business Services (KIBS)

Figure 12.1b indicates that KIBS are most important in the largest core cities, Manchester, Birmingham and Leeds (>12.5 % of total employment), although Bristol (17.1%) has the highest share of all. Their surrounding regions have lower KIBS shares, but still higher than the hinterlands of the other cities (>7%). We have seen that KIBS lead growth in and around Liverpool and Newcastle, and in the city of Leeds, and that slow knowledge-intensive growth in Birmingham was compensated for by rapid expansion in its hinterland with the same true for KIBS of Bristol (Figure 12.2b).

The composition of KIBS in each city throws some light onto the nature of recent changes. Even though dominated nationally by London, the **financial services** are the most important (Table 12.2), especially financial intermediation (SIC 65: e.g. banking, building societies finance houses, investment trusts, etc) and insurance (life and non-life), supplemented by auxiliary activities such as fund management, stock broking and underwriting. Their hinterland regions also support significant numbers, especially around Leeds (e.g. in Bradford and Halifax) and Liverpool (e.g. Southport). Functionally, however, most of these activities serve local retail markets. Where they are most concentrated, in Leeds, Manchester, Birmingham, Liverpool and Bristol, they may also serve wider regional, and possibly national and international markets.[11] A further index of this quality may be the presence of the '**auxiliary**' financial activities (SIC 67). These are nationally particularly dominated by London, but are best represented among the core cities in Manchester, with 6,700 employees in 2002, and Bristol, with almost 5,000.

---

11 Table 12.2 highlights the 2002 core city employment share of KIBS and creative activities, **shaded** for those with more than London's share, and **bold** for those with more than half London's share (i.e. the core city average for all knowledge-intensive activities).

Table 12.2   KIBS and creative services: core city employment share compared with Central London

Bold: a. > half central London's share; b. +/- 15% change; Shaded: =/> central London's share: +/- 30% change

SIC categories:	Manchester a. Share/ London	Manchester b. % Change 1998-2002	Birmingham a. Share/ London	Birmingham b. % Change 1998-2002	Leeds a. Share/ London	Leeds b. % Change 1998-2002	Liverpool a. Share/ London	Liverpool b. % Change 1998-2002
Knowledge-Intensive Business Services								
65 : Financial intermediation, etc	0.33	25.9	0.46	8.6	0.59	30.7	0.35	26.1
6601 Life insurance	1.83	-15.1	1.00	-14.0	0.67	4.5	2.50	96.4
6603: Non-life insurance	0.78	-8.9	0.67	-19.8	1.56	52.3	1.56	31.3
67: Auxiliary financial intermediation	0.26	62.3	0.19	-0.3	0.19	7.0	0.15	400.6
7220:Software consultancy and supply**	0.65	46.6	0.65	41.8	0.59	54.0	0.29	-1.9
7411 : Legal activities	0.50	31.7	0.45	33.9	0.37	52.3	0.50	25.9
7412; Accountancy/book-keeping, etc	0.45	0.7	0.59	-14.6	0.41	-5.2	0.27	-8.5
7414: Business/management consulting	0.29	44.0	0.25	32.2	0.21	8.8	0.29	400.6
7420: Architectural/Engineering**	0.74	-6.9	0.63	5.3	0.53	15.4	0.53	61.6
'Creative' sectors (excluding above categories)**								
221: Publishing	0.13	-24.8	0.17	-15.8	0.22	5.4	0.17	-4.8
744: Advertising	0.46	18.0	0.23	3.8	0.23	16.5	0.08	-17.2
9220: Radio and TV activities	0.45	118.2	0.27	5.7	0.27	14.5	0.27	-45.8
9231: Artistic and literary creation etc.	0.30	5.6	0.20	0.8	0.20	-2.7	0.20	-9.2
9232: Operation of arts facilities	1.00	369.0	0.50	334.7	-	-	0.50	319.8

	Sheffield		Newcastle		Nottingham		Bristol	
	a. Share/ London	b. % Change 1998-2002	a. Share/ London	b. % Change 1998-2002	a. Share/ London	b. % Change 1998-2002	a. Share/ London	b. % Change 1998-2002
Knowledge-Intensive Business Services								
65 : Financial intermediation, etc	0.41	32.3	0.27	14.1	0.31	25.4	0.53	-5.0
6601 Life insurance	1.33	3.7	0.50	-21.8	0.17	-75.4	3.67	54.3
6603: Non-life insurance	0.11	-32.5	0.33	-28.2	0.22	34.2	1.22	2.0
67: Auxiliary financial intermediation	0.19	93.8	0.11	-5.4	0.11	26.6	0.40	10.6
7220:Software consultancy and supply**	0.29	-3.1	0.59	58.1	0.65	-2.9	0.65	32.0
7411 : Legal activities	0.32	12.2	0.34	27.0	0.34	8.6	0.50	10.9
7412; Accountancy/book-keeping, etc	0.32	2.5	0.18	82.4	0.36	-11.3	0.45	0.1
7414: Business/management consulting	0.13	-25.1	0.33	8.9	0.13	13.3	0.25	55.3
7420: Architectural/Engineering**	0.37	-26.3	0.68	6.1	0.58	-2.9	0.79	-13.7
'Creative' sectors (excluding above categories)**								
221: Publishing	0.13	-18.8	0.26	17.2	0.09	-19.4	0.30	69.7
744: Advertising	0.23	89.7	0.31	-5.2	0.08	-27.0	0.38	38.1
9220: Radio and TV activities	0.09	49.9	0.18	-1.2	0.27	-20.2	0.45	-44.6
9231: Artistic and literary creation etc.	0.30	47.4	0.10	1.7	0.20	53.9	0.30	5.5
9232: Operation of arts facilities	-	-	-	-9.6	-	-	-	-

< 50 employees: - : Losses in *Italics*

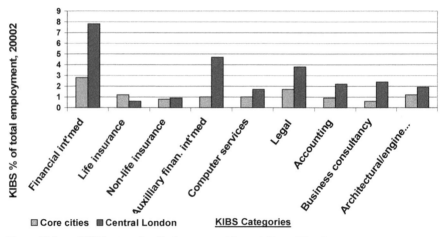

Figure 12.4a   **KIBS categories, % share: core cities and Central London**

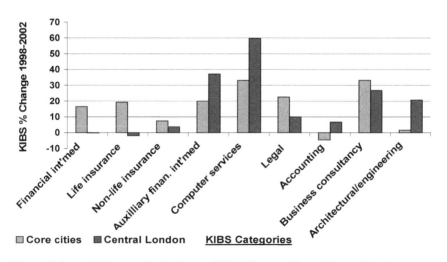

Figure 12.4b   **KIBS categories, % change, 1998-2002: core cities and Central London**

**Figure 12.4   KIBS categories, % share and % change, 1998–2002:
All core cities and Central London**

The greatest share of core city financial services compared to central London, however, is in **Life Insurance** (6601), although everywhere employing many fewer than banking. In fact, several core cities possess higher employment shares than central London (shaded in column a, Table 12.2), including Bristol, Liverpool, Sheffield and Manchester. These are augmented in Leeds, Liverpool and Bristol by similar concentrations of **non-life insurance** activity (6603), although the distinction is increasingly blurred as large companies engage in both. Like financial intermediation, life insurance showed striking core city growth between 1998 and 2002, contrasting with the sector's stagnating numbers in central London (Figure 12.4b). The two southern cities, Birmingham and especially Bristol, followed London in not expanding banking-related employment. Insurance trends were more patchy; growing in Leeds, Liverpool and Bristol, but losing significant numbers in Manchester, Birmingham and Newcastle. These outcomes no doubt reflect local corporate restructuring in UK financial services, overlain by the impacts of the London-based cyclical recession after 2000.

Paradoxically, core city growth during this period tends to reinforce the impression that the regional financial sectors are functionally different from London-centred international activities. While the latter suffered a setback after 2000, credit-driven consumer demand remained strong in the UK and was generally recognized to be protecting the economy from recession. The orientation of most core city financial services to domestic markets thus supported their growth. In contrast, financial services in London generally suffered from the recession, but the more expertise-intensive, internationally-orientated 'auxilliary' services (SIC 67) continued to favour the capital. Already almost five times more concentrated there than in the core cities, their central London employment grew at almost twice the core city rate between 1998 and 2002 (Figure 12.4b). As well as this difference in market orientation, technology-driven cost squeezes on routine functions also affected London-based financial employment. This benefited the northern core cities, with their labor availability and lower operating costs. Meanwhile, the concentration of international strategic financial functions in London appears as strong as ever. This pattern, recognized since the 1980s (Leyshon and Thrift 1989; 1997), continues to raise questions about the quality and security of the KIBS growth in core cities. Some of this employment may be particularly vulnerable to ICT-based developments, such as web-based banking and insurance, and competition from offshore locations.

Technical services provide the second, and possibly most strategically significant KIBS focus in the core cities and their regions, represented in Table 12.2 by **engineering and architectural consultancy** (7420). In all except Sheffield, their employment share is closer to central London's than the KIBS average, especially Bristol and Manchester. It is also high in the Bristol and Nottingham regions, associated with aerospace manufacturing. Such specialist technical functions have often grown out of local industrial strengths, serving regional manufacturing, commercial and public sector markets. They often serve UK-wide markets, and some have international reputations, a potential focus for future core city knowledge-intensive developments. Unfortunately, their core city employment levels were virtually stagnant between 1998 and 2002, in contrast again to continuing strong growth in central London (Figure 12.4b). There were significant employment losses in Manchester, Nottingham, and especially Sheffield and Bristol. The only core cities to expand these functions were Liverpool and Leeds.

The most expansive KIBS sector is another technical advisory service: **computer software consultancy**, including systems (7220). Its representation in most core cities was again relatively high, except for Liverpool and Sheffield. Once more, however, their growth was disappointing (Figure 12.4b). Although employment expanded by one third (33.1%) in four years, this was much lower than central London (60%), even with its recession. Manchester, Leeds, Birmingham and Newcastle almost reached London levels of growth, but Liverpool, Sheffield and Nottingham seemed actually to lose jobs. Computer service employment was also important in the surrounding regions, however, especially around Bristol, and Birmingham (in Solihull, north Warwickshire and nearby parts of Staffordshire), as well as Newcastle.

The KIBS profile of all the core cities includes a significant base of **legal activities** (7411), especially in Manchester, Liverpool, Bristol and Birmingham, which generally expanded more rapidly than London's between 1998 and 2002. **Accountancy** (7412) was relatively well represented in Birmingham, but this was the only core city where its employment share was over half that in central London. The numbers were generally stagnant, in line with national trends. The quality of these professional services, especially in offering local business services that compare in quality and international reach to London's, is an important element in regional KIBS provision, but again much of their work is for individuals rather than businesses. A more significant index of the quality of locally-based business support is the presence of **business and management consultancy** (7414). These specialist business services remain heavily concentrated into the capital. No core city's employment rises above one third of central London's share. Manchester, Birmingham, Liverpool and Bristol showed quite strong growth from a low base, but no sign of significantly redressing the current regional imbalance. There are more localized specializations, at least compared with the low core cities' average, include **data processing** in the Nottingham region and Newcastle; **market research** in Leeds and Liverpool; and **real estate** management in Manchester. There is also higher than average **advertising** employment in Manchester and Bristol, but generally most core cities lack nationally significant employment in most of these functions.

The 'Creative' Sectors

'Creative' services grow through the imaginative development of computer and other technical capabilities, marketing skills and artistic originality. In their commercial manifestations they interact with KIBS functions, so that the largest 'creative' employment numbers are found in software development (with other computing) and architectural/engineering consultancy, already discussed. Advertising is here classified as 'creative', although it is also a 'business service'. These unavoidable definitional overlaps indicate the importance in a knowledge-based economy of functions that both create new ideas and approaches, and link them to production methods and employment generation. They may also be important in attuning productive output to social and cultural trends, sometimes setting such trends, especially through the modern media.

Only Bristol and Manchester among the core cities approach half of central London's share of creative employment (Figure 12.1c), but these activities are also well represented in the wider Bristol region. Newcastle and Nottingham had comparatively high shares, and Liverpool and Sheffield the lowest. There were also divergent trends between 1998 and 2002 (Figure 12.2c). Leeds and Newcastle expanded faster than central London, while Birmingham, Nottingham Sheffield, and especially Bristol fell behind. For Bristol and Birmingham, their surrounding regions seemed to compensate, as we have seen mainly through computer software and technical services.

The comparatively disappointing employment trajectory of architectural and engineering services in some core cities is doubly significant, since they are sources of both business expertise and creativity. Their creative importance is nevertheless even more difficult to represent through employment data than their business service role. Much more detailed enquiries are needed to pursue such issues. These functions are also often equally important for employment in surrounding regions. The 33,000 workers in Manchester and its region certainly suggest a critical mass of urban-based expertise as a focus for growth. The comparable figures for Birmingham and its region are 27,000, and for Leeds 21,000. Quality is even more critical among the smaller cities. The Nottingham (18,000) and Bristol (17,000) regions depend heavily on their high technology aerospace activities. For Newcastle (14,000), Liverpool (12,000) and Sheffield (10,000), the focus seems more diffuse.

Perhaps the most striking development during the period was the expansion of **arts facilities** in Manchester, Birmingham, Liverpool, Sheffield and Nottingham (Table 12.2). Although from a low initial level, this reflects investment in cultural infrastructure through National Lottery-generated funding, explicitly to stimulate wider economic urban economic revival. Otherwise, much of the private-sector creative involvement is focused in Manchester and Bristol. **Publishing** employment is generally weak and declining, dominated by technological developments in newspaper production. There are also local clusters of **Radio and TV activities**, often regarded as important in projecting an alternative to the London-dominated media, but numbers and growth were significant only in Manchester (around 3,000 employees). **Advertising** is nationally London-dominated, and again only the Manchester numbers (3,000), and perhaps Bristol, suggest any distinctive local capacity.

High Technology

In general, conventionally defined high technology activities are not very important in the core cities, and even less in London. They are found more in some hinterland regions, especially around Bristol, Nottingham and Manchester (Figure 12.1d).

Computer hardware and service provision provide the most widespread and rapidly growing high technology functions. With software consultancy, they offer the most potentially creative sources of technologically-based innovation in the wider core city regions, in association with other manufacturing and services. The pattern of change is nevertheless one of overall stagnation, with local small-scale growth in some CCRs (e.g. Liverpool, Bristol), balanced by decline elsewhere (Sheffield, Newcastle, Nottingham) (Figure 12.2d).

Employment in nominally high technology functions is, of course, often most at risk because of the intensity of competition in such activities, often from abroad, and the quite different location requirements of the innovation and production stages of development. The mere presence of high technology functions is not enough – they have to be successful, supported locally by corporate capital and integrated with other service aspects of the regional economy. Even in London, growth has been largely focused on market-driven computer-related production and services, with other activities, such as pharmaceuticals, engineering, and electrical equipment, stagnant or declining. Nationally, of course, South East England outside both London and the core cities dominates the pattern of technologically innovative activities.

Higher Education

Research and teaching in higher education institutions represent major public investments in scientific and other intellectual innovation and advanced workforce training. Many studies have been undertaken into the urban and regional impacts of such investment (Goddard 1997; Chatterton 1997), but little is known about these in the English core cities or London. The largest core city concentration of higher education employment is in Leeds and its region (28,000), followed by Manchester (24,000) and Birmingham (20,000). It occupies the highest proportions of the regional workforce in Leeds, as well as Newcastle, followed by Liverpool (Figure 12.1e). Highly rated research departments (awarded 5 and 5* assessments in 2002 national Research Assessment Exercise) are relatively concentrated in Manchester, Nottingham and Sheffield, with the share of science and engineering departments amongst these highest in Leeds and Liverpool, and lowest in Birmingham and Manchester (and London). All the core cities therefore possess substantial resources of high quality scientific research, and this may be of special significance in Leeds and Liverpool.

Knowledge-Intensive Activities and the Futures of Core Cities

The introduction to this chapter argued that business, technological and creative knowledge intensive functions, whether available within organizations or from specialist providers, locally or from outside, must be the basis for successful regional, and especially urban, adaptability to economic change. The revival of the English core cities thus depends on such functions to develop new suites of competitive, tradable activities around which key actors, institutions and relationships may focus. The cities face particular problems because of the historic and growing dominance of the knowledge-intensive service nexus based in and around London. Recent national economic policies have tended to reinforce this by supporting a deregulated private, and centralized public knowledge economy. The growth of knowledge-intensive functions in the core cities has nevertheless been seen as a sign of economic recovery, especially since the late 1990s, when London employment growth slowed.

The evidence reviewed here, however, especially for critical sectors such as the auxiliary financial, business, architectural, engineering, and even computer services, suggest little weakening in London's dominance over UK internationally tradable services. In some cases, it seems that opportunities are being lost to exploit the core cities' inheritance of technical and commercial expertise. Instead, up to 2002 their service employment growth seemed to depend more on the debt-based dynamism of UK consumer markets, and their attractions for cost-sensitive, domestic or routine administrative functions. This certainly appears to be the case for the smaller northern cities such as Newcastle, Sheffield and Liverpool. The more dominant northern cities, Manchester and Leeds, may have developed more sophisticated functions to serve wider markets, at least within the UK (see Britton et al. 2004, for a detailed study of Manchester). Bristol, after its earlier rapid expansion, apparently followed London in suffering market recession and cost-based rationalization. The Midlands cities, Birmingham and Nottingham, fared worst, affected by recession but without the compensating attractions of northern cities. There was also marked decentralization to their surrounding regions around Birmingham, Bristol and Liverpool.

As has been emphasized, this type of employment analysis cannot indicate the quality of work carried out by knowledge-intensive functions in the core cities, or their potential for innovative change and growth. Even if this were possible, however, it would probably not show the core cities in a favourable light. On the contrary, London's net dominance of national commercial knowledge employment, and its export-earning capacity, may well have grown, as routine functions have declined or moved elsewhere. This trend may even have been accelerated by the recession of 2000–2003. In such a situation, it needs to be recognized that disparities in their size and historical roles will always place smaller UK cities in a poor light compared with London. It is more fruitful for economic development indicators to be compared between the core city regions themselves and, as we have seen, ideally with similar regions elsewhere in Europe.

Focusing on KIBS strengths, for example, in all urban regions these essentially depend on the success of other regional businesses, providing a means of supporting their continuing success. Although private sector employment in the core cities is increasingly dominated by businesses owned elsewhere, they support many significant plants and offices with international connections, similar to those identified by O'Connor for cities in the US and Australia (O'Connor 2003). The relationships at such sites between in-house expertise and specialist KIBS should be a starting point for fostering local knowledge-based economic development. Then there may be regionally-based businesses with growth potential. If these are to prosper, what are their knowledge-intensive requirements, and how can they be satisfied, from local or outside sources? Business advisory services are sometimes offered by public bodies as a substitute for commercial KIBS, but the public sector in the UK is also a major and growing KIBS market. Is there scope for creating a critical mass of local demand for commercial KIBS, combining the needs of local businesses and the public sector? The inherently inter-dependent basis of regional KIBS success, often ignored in sector-based economic commentaries, thus presents both problems and potential for policies. These should not focus on narrow 'target sectors', however, but encompass the quality of all the local knowledge-intensive inputs likely to support economic development.

Another aspect of core city KIBS potential is the access their businesses have to London-based KIBS. Seen as a purely competitive relationship, this may appear to undermine local KIBS development. But it should also be a major asset for UK-based companies generally, especially when operating in international markets. There is also evidence that, where there is sufficient regional demand, KIBS firms with particular specialist and 'local' knowledge will develop to serve it (O'Farrell et al. 1992; Wood et al. 1993). Competition from London-based KIBS may even encourage the quality of regional KIBS firms, not least through niche technical, organizational or marketing specialization, in turn providing a basis for serving wider markets (O'Farrell, et al. 1996). They are also often able to offer lower prices for comparable levels of expertise. So, national and regionally-based KIBS in Manchester and Leeds, for example, have some advantages in serving the needs of northern England, including the multinational and other export-based companies based there. At the same time, over-concentration in the South East has encouraged some KIBS to disperse to other regions, especially in and around Birmingham, Nottingham and Bristol. Thus, either by responding to and supporting regional potential, or by spinning-off London's strengths, the core city regions can potentially develop various KIBS-related roles, including in overseas markets, associated with major clients. We need cross-sectoral policies that are aware of these evolving spatial relationships within the national knowledge-intensive economy.

Conclusions: Knowledge-Intensive Services as the Focus for Core City Development

Core city region knowledge-based policies should therefore be assembled from the following components:

1) Their KIBS employment is dominated by the **financial services**, but these cannot be relied upon to grow significantly in the future. Compared to London, any truly 'knowledge-intensive', tradable functions are submerged in domestic consumer and business financial activities. The presence of the 'auxiliary' services (SIC Division 67) provides one index of quality, but remain small in the core cities and dominated by London. **Life insurance** is the only KIBS activity occupying a higher core city employment share

than London. But insurance is traditionally provincially-based, and again largely involved in the routine management of UK life and non-life funds. It is not internationally innovative and, as for financial services generally, corporate restructuring and ICT-driven rationalization, some favouring offshore locations, threaten future employment. Certain financial services in the larger cities do offer specialist expertise to regional business and may serve wider national and even international markets. More needs to be known about these and the conditions supporting their development, including the scale and quality of regional demand, international connectivity, and ability to attract key staff.

2) The most distinctive strength of the core cities lies in the various types of **'technical consultancy'**, including construction, architectural, engineering and computer systems and software. Some of these have grown out of the cities' traditional industrial, commercial and construction strengths, and include high proportions of 'knowledge-intensive' staff. Such services can play a multiple economic role, combining commercial, technological and creative forms of knowledge to serve diverse markets, in manufacturing, construction, transportation and trading, commercial and consumer services, and the public sector. They also offer a relatively secure basis for technologically advanced work, not necessarily tied to specific production processes, products, markets or clients. Recent core city employment trends nevertheless offer little evidence that these opportunities are being grasped, at least enough to counter the attraction of London as a location for international businesses. A more concerted effort is required to promote the core cities as centers of excellence in technical knowledge and practice. This also requires a close association with high technology regional manufacturing, construction, business and consumer services, and the public sector including universities. The legacy of the core cities' economic past does not currently seem to be being promoted sufficiently to resist the drift of such tradable functions to London.

3) **Legal and accountancy and real estate** functions will continue to play a significant employment role in the larger core cities. Although primarily orientated to regional clients, these increasingly expect access to national and international (especially EU) standards of legal, accountancy, property management and other expertise. Core cities thus need to sustain the quality of such business, both to retain employment and ensure that the regional business environment is as supportive as other cities, including London.

4) There may be potential for development among the **'creative'** sectors; the media activities of TV, radio and film, the arts, and advertising. The level of activity is still modest, however, and subject to many unpredictable creative, technological, corporate and policy developments. The quality of regional demand will be most significant in sustaining them, but recent growth seems to reflect some benefits from the cultural revival of the core city regions, as well as regional devolution by dominant agencies such as the BBC.

5) The future of **technological innovation** in the core cities depends on the focus, technical sophistication and corporate context of local manufacturing and service businesses. Even successful innovative outcomes, however, do not guarantee beneficial employment impacts. Some core city regions have significant high technology manufacturing, including aerospace, computer hardware, electrical equipment, precision instruments, and pharmaceuticals. As has been suggested, the most creative and secure high technology employment depends on local associations between the corporate and R&D functions of manufacturing firms and expert technical, including computer consultancies. These may also spawn growth-orientated SMEs, but must also involve active exchanges with national and international technological, business and market expertise.

6) Such strategies must also benefit in various ways from one of the most distinctive strengths of the core cities, in quality **higher education**. Close regional links need to be fostered with the innovative potential created by its scale and quality of research, through both technological development and economic and social intelligence. National higher education policies now encourage such regional spin-off, but these need to be matched by local agencies. This may also encourage the retention of high quality local graduates.

7) Exploiting the **knowledge-based potential** of the core cities also requires strategies combining it with that of their hinterland regions. Many KIBS are likely to remain in the cities, but the critical mass of demand required to support specialist knowledge-intensive growth should, at the least, be based on their city-regions. Their technical and computer-based creative capacities are already relatively dispersed into the surrounding regions.

The English core cities remain major regional centres with great intellectual, economic and cultural resources. They appear to have enjoyed some renaissance over the past twenty years, at least as indicated by city centre revivals and associated real estate development. Major new cultural, entertainment, sporting and retailing facilities have promoted a veneer of wellbeing. But, when their engagement with the knowledge-intensive basis of the UK's international competitiveness is examined, they appear to be making only limited progress. There also seems to be a real danger of further marginalization, for example, if financial services employment declines further, or the regions fail to build a technical service basis out of their industrial traditions. To avoid this outcome, a greater capacity will be required to co-ordinate the promotion and development of KIBS and creative services, innovative manufacturing, and higher education. The examples of London, through its international position and scale of activity, and of the continental non-capital cities, with their localized capacity to manage development, demonstrate the competition the core cities face.

Exploiting the complementary assets of the cities and their surrounding regions also needs to extend to coordination between adjacent cities, such as Manchester and

Liverpool, Leeds and Sheffield, and even the 'trans-Pennine' potential of northern England more broadly.[12] Some regions, such as Manchester, Leeds and Bristol possess a stronger knowledge economy potential than others, each in different ways. All would nevertheless benefit from a policy focus on the knowledge-intensive economy as a whole, rather than on manufacturing compared with services; the private rather than the public sector; high technology innovation rather than adoption; or specialist innovative 'winners' rather than active expertise exchange supporting all types of new ideas. In this, the core city studies concluded that the experience of the European non-capital cities indicates the direction in which the core cities might move to reverse their trajectory of longer-term decline in the era of the 'new knowledge economy'. It is equally clear, however, that this would require significant changes, beyond the current 'core cities' rhetoric, in long-established UK government attitudes to allowing greater economic localism.

References

Allen, J. (1992), 'Services and The UK Space Economy: Regionalization and Economic Dislocation', *Transaction, Institute Of British Geographers*, NS 17, 292–305.
Amin, A. and Thrift, N. (1992), 'Neo-Marshallian Nodes in Global Networks', *International Journal Of Urban And Regional Research* 16, 571–87.
Bathelt, H. and Boggs, J.S. (2003), 'Towards a Reconceptualization of Regional Development Paths: Is Leipzig's Media Cluster a Continuation of or Rupture with The Past?', *Economic Geography* 79, 265–93.
Bathelt, H. and Glückler, J. (2003), 'Towards a Relational Economic Geography' *Journal of Economic Geography* 3, 117–44.
Beyers, W.B. (2002), 'Services and the New Economy: Elements of a Research Agenda', *Journal Of Economic Geography* 2, 1–29.
Boggs, J.S. and Rantisi, N.M. (2003), 'The "Relational Turn" in Economic Geography', *Journal Of Economic Geography* 3, 109–16.
Britton, J., Halfpenny, P., Devine, F. and Mellor, R. (2004), 'The Future of Regional Cities in the Information Age: The Impact of Information Technology on Manchester's Financial And Business Service Sector', *Sociology*, 38(4), 795–814.
Butchard, R.L. (1987), 'A New Definition of High Technology Industry', *Economic Review* 400, 82–8.
Chatterton, P. (1997), *The Economic Impact of the University of Bristol on Its Region*, University of Bristol.
Department of Trade and Industry (2003), *Innovation Report: Competing in the Global Economy: The Innovation Challenge*, December, Her Majesty's Stationery Office.

12 In 2004, the Regional Development Agencies for the North East, North West and Yorkshire established a task group to examine strategies for a 'Northern Growth Corridor' to coordinate growth assets across the three regions.

European Commission (2002), European Innovation Scoreboard, *Technical Paper No 3, Regional Innovation Performances*. Enterprise & Industry Directorate General, Innovation Policy Development Unit.

Goddard, J. (1997), *Universities and Regional Development*, University of Newcastle upon Tyne, School of Management/ Centre for Urban and Regional Development.

Harrison, B. (1992), 'Industrial Districts: Old Wine in New Bottles?' *Regional Studies* 26, 469–83.

Le Galès, P. (2001), *European Cities: Social Conflict and Governance* (Oxford: Oxford University Press).

Leyshon, A. and Thrift, N. (1989), 'South Goes North? The Rise of British Provincial Financial Centres', in Lewis, J. and Townsend, A. (eds), *The North-South Divide: Regional Change in Britain in the 1980s* (London: Paul Chapman), 114–56.

Leyshon, A. and Thrift, N. (1997), *Money/Space: The Geographies of Monetary Transformation* (London: Routledge), Chapters 4–7.

Markusen, A. (1996), 'Sticky Places in Slippery Space: A Typology of Industrial Districts', *Economic Geography* 72, 293–313.

Martin, R. and Sunley, P. (2003), 'Deconstructing Clusters: Chaotic Concept Or Policy Panacea?', *Journal of Economic Geography* 3: 5–35.

O'Connor, K. (2003), 'Rethinking Globalization and Urban Development: The Fortunes of Second-Ranked Cities', *Globalization and World Cities*, Loughborough University, Research Bulletin 118, August, http://www.Lboro.Ac.Uk/Gawc/Publicat.Html,

Office of the Deputy Prime Minister (2004), *Competitive European Cities: Where Do the Core Cities Stand?* Report By Parkinson, M., Hutching, M., Simmie, J., Clark, G. and Verdonk, H., HMSO, January.

O'Farrell, P.N., Hitchens, D.M. and Moffat, L.A.R. (1992), 'The Competitiveness of Business Service Firms: A Matched Comparison Between Scotland and the SE of England', *Regional Studies* 26, 519–34.

O'Farrell, P., Wood, P. and Zheng, J. (1996), 'Internationalization in Business Services: An Inter-Regional Analysis', *Regional Studies* 30, 101–18.

Porter, M.E. (1998), 'Clusters and the New Economics of Competitiveness', *Harvard Business Review*, December, 77–90.

Porter, M.E. (2000), 'Locations, Clusters and Company Strategy', in Clark, G.L., Feldman, M.P. and Gertler, M.S. (eds), *The Oxford Handbook Of Economic Geography* (Oxford: Oxford University Press), 253–74.

Scott, A.J. (1988), New *Industrial Spaces: Flexible Production Organization and Regional Development in North America and Western Europe* (London: Pion).

Scott, A.J. (1998), *Regions and the World Economy* (Oxford: Oxford University Press).

Scott, A.J. (ed.) (2001), *Global City Regions: Trends, Theory and Policy* (Oxford: Oxford University Press).

Simmie, J., Wood, P. and Sennett, J. (2002), 'Innovation and Clustering in the London Metropolitan Region', in Begg, I. (ed.), *Urban Competitiveness* (Bristol: The Policy Press), 161–90.

Simmie, J., Siino, C., Zuliana, J-M., Jalabert, G. and Strambach, S. (2004) 'Local Innovation System Governance and Performance: A Comparative Analysis of Oxfordshire, Stuttgart and Toulouse', *International Journal of Technology Management* 28, 534–59.

Storper, M. (1997), *The Regional World: Territorial Development in a Global Economy* (New York: Guilford).

Sturgeon, T.J. (2003), 'What Really Goes on in Silicon Valley? Spatial Dispersal and Clustering in Modular Production Networks', *Journal of Economic Geography* 3, 199–225.

Swyngedouw, E. (1996) 'Neither Global or Local: "Glocalization" and the Politics of Scale', in Cox, K. (ed.), *Spaces of Globalization: Reasserting the Power of the Local* (New York: Guilford), 137–66.

Wood, P. (2002a), 'The United Kingdom: Knowledge-Intensive Services in a Restructuring Economy', in Wood, P. (ed.), *Consultancy And Innovation* (London: Routledge), 175–208.

Wood, P. (2002b), 'Services and the "New Economy": An Elaboration', *Journal of Economic Geography* 2, 109–114.

Wood, P. (2006), 'Urban Development and Knowledge-Intensive Business Services: Too Many Unanswered Questions?' *Growth and Change*, 37, 335–61.

Wood, P., Bryson, J. and Keeble, D. (1993), 'Regional Patterns of Small Firm Development in the Business Services: Evidence from the UK', *Environment and Planning A* 25, 677–700.

Yeung, W.Y. (1998), 'Capital, State and Place: Contesting the Borderless World', *Transactions, Institute of British Geographers* 23, 291–310.

Appendix 12.A   SIC definitions of knowledge-intensive functions at 4-digit level

1) Knowledge-intensive business services (KIBS)

i) The wholesale **financial** services:

65:	Financial intermediation
6601/2:	Life Insurance/Pension funding
6603:	Non-life insurance
67:	Auxiliary to financial intermediation

ii) Business-orientated **property development and management** (excluding estate agencies):

7011:	Development and selling real estate
7032:	Management of real estate

iii) Knowledge-intensive **'Other business activities'**, excluding functions such as catering, security, cleaning, packaging, secretarial agencies and labor recruitment.

7210:	Hardware consultancy
7220:	Software consultancy
7230:	Data processing
7240:	Data base activity
7411:	Legal activities
7412:	Accounting, book-keeping, etc.
7413:	Market research/public opinion polling
7414:	Business/management consultancy
7415:	Management activities; holding companies.
7420:	Architectural/engineering activities
7430:	Technical testing and analysis
7440:	Advertising

2) The 'creative industries'

(Technical and artistic, some overlapping with KIBS)

221:	Publishing
72:	Computing and related activities
742:	Architectural/engineering activities
744:	Advertising
7481:	Photographic activities
9211:	Motion picture and video production
9220:	Radio and television activities
9231:	Artistic and literary creation, etc.
9232:	Operation of arts facilities

3) High technology sectors (Adopting definition of Butchard 1987):

2416	Manufacture of: plastics
2417	synthetic rubber
2441	pharmaceuticals
2442	pharmaceutical preparations
3001	office machinery
3002	computers, etc.
3110	electric motors/generators
3120	electrical distribution apparatus
3162	electrical equipment nec.
3210	electronic valves, etc.
3220	TV/radio transmitters, etc.
3310	medical/surgical equipment
3320	measuring instruments, etc.
3330	industrial process control equipment
3340	optical instruments, etc
3530	aircraft and spacecraft
7210	Hardware consultancy
7260	Other computer related activities
73	Research and development

4) Higher education   803: Higher Education

Appendix 12.B   Area definitions and population, '000s, 2001

Local authority areas

Manchester City:	Manchester, Trafford, Salford:	858
Manchester region:	Bolton, Bury, High Peak, Macclesfield, Oldham,	
	Rochdale, Rossendale, Stockport., Tameside, Wigan:	2,789
TOTAL		**3,647**
Birmingham:	Birmingham:	976
Birmingham region:	Bromsgrove, Cannock Chase, Dudley, Lichfield,	
	North Warwickshire, Redditch, Sandwell, Solihull,	
	South Staffordshire, Tamworth, Walsall,	
	Wolverhampton, Wyre Forest:	2,927
TOTAL		**3,903**
Leeds:	Leeds:	715
Leeds region:	Bradford, Calderdale, Harrogate, Kirklees,	
	Selby, Wakefield, York:	2,489
TOTAL		**3,204**
Liverpool:	Liverpool:	439
Liverpool region:	Halton, Knowsley, Sefton, St Helens, West Lancashire,	
	Wirral:	1,588
TOTAL		**2,027**
Sheffield:	Sheffield:	513
Sheffield region:	Barnsley, Bassetlaw, Bolsover, Chesterfield,	
	Doncaster, NE Derbyshire, Rotherham:	1,642
TOTAL		**2,155**
Newcastle:	Gateshead, Newcastle upon Tyne:	451
Newcastle region:	Alnwick, Blyth Valley, Castle Morpeth, Derwentside,	
	Durham, North Tyneside, South Tyneside, Sunderland,	
	Tynedale, Wansbeck:	1,530
TOTAL		**1,721**
Nottingham:	Broxtowe Gedling, Nottingham:	485
Nottingham region:	Ashfield, Mansfield, Erewash, Derby, Amber valley,	
	Rushcliffe, South Derbyshire:	1,332
TOTAL		**1,443**
Bristol:	Bristol:	381
Bristol region:	Bath and N E Somerset, Mendip, N Somerset,	
	South Gloucestershire:	1,091
TOTAL		**1,472**
Central London:	City, Westminster, Tower Hamlets, Kensington & Chelsea,	
	Camden, Islington, Southwork, Lambeth:	1,432
'London region':	Rest of Greater London:	5,769
TOTAL		**7,192**

Chapter 13

Innovation Activities of KIBS Companies and Spatial Proximity: Some Empirical Findings from Finnish New Media and Software Companies

Jari Kolehmainen

Services and Innovation

Companies' success is increasingly determined by their innovation and learning. In knowledge-intensive sectors such as the new media and software business, this challenge is even bigger than in the more traditional industrial and service sectors. The very concept of innovation has changed. Today innovation is increasingly seen as a 'circular' or 'recursive' process instead of the old view of innovation as "commercialized invention based on technological or scientific knowledge". In addition to this "cascade model", the so-called "market-pull" innovation model represents also linear innovation thinking, whereas the "recursive" innovation model stresses the versatile feedback mechanisms and interactive relationships involving producers (companies), product users, scientific and technical research, development activities, and supporting infrastructure (Schienstock and Hämäläinen 2001). It is also a model of continuous learning, in which the actors in different arenas learn from each other in interactive innovation processes. This means that many actors are involved in a single innovation process, and it can be triggered by many causes (cf. Miettinen et al. 1999). Therefore both explicit inter-organizational innovation networks and social linkages have become crucially important (e.g. Faulkner 1995).

The current understanding has broadened the scope of innovation activities. In addition to major, invention-based technological innovation, attention is drawn to more incremental improvement and development of products, processes, collaborative relationships, brand, delivery channels, etc. The linear innovation models were developed largely from experiences in manufacturing and the service sector was to some extent neglected. However, increasing attention has been paid to the services, because they play crucial roles in the modern economy. Manufacturing industries have become less labor intensive because of technological change, such as the development of automation and adoption of new information and communication technology. Correspondingly, the role of services in the economy has grown, especially services referred to as knowledge-intensive services (KIS) and knowledge-intensive business services (KIBS). These services are important for the development of manufacturing industries, because they offer specialized competences to companies that have focused strictly on their core competences. At the same time they are an important element in the national and regional innovation systems. As Kautonen (2001) points out, the KIBS have many roles in the innovation system or in the innovation environment. They can act as carriers, facilitators, or as even sources of innovation for their customer companies. Most often KIBS companies affect indirectly their customers' innovation processes by providing services that enable the customers to carry out their innovative activities.

There are many definitions for KIBS because it is very hard to define these services exactly. Nählinder (2002, 4) has found three characteristics that are present in most definitions. First, KIBS produce information or service products which generate knowledge for other companies and these services are based on human capital. One of the major functions of KIBS is to foster "knowledge development" in other parts of the economy. Secondly, KIBS companies are deliverers of service products that are based on new or emerging technologies, such as ICT. Thirdly, KIBS companies design and provide the service products usually in close interaction with the users.

KIBS companies play a large role in the innovation activities of their customers, but they face also themselves the same innovation challenges common to service providers. Kuusisto and Meyer (2003) suggest that many service innovations are based on the possibilities enabled by information technology or market changes related to globalization and regulation. They also have detected an emerging tendency towards more deliberate attempts to innovate in service, which are targeted to improve the cost efficiency and quality of service production and products as well as to develop new service concepts. Generally speaking, the innovation activities of the service companies are more heterogeneous than those of manufacturing companies (Leiponen 2000). As Kuusisto and Meyer (2003) note, a significant portion of innovation patterns in services is 'soft', or non-technological, even in product and process innovations. Technology-related services are naturally more technologically oriented also in innovation activities than service companies. Piirainen et al. (2000) have studied the innovation activities of KIBS companies in Tampere Central Region and they have found some typical features (see also Kautonen et al. 2002):

- Service development is integrated with customer projects.
- Service development is more or less unorganized: no specialized unit is formed within a company, although this is typical for an average small companies producing tangible products as well.
- Development activities are usually carried out side by side with normal daily routines and in ad hoc teams.

- Companies' own internal processes of production are developed during "silent" periods, i.e. when there is time left from the customer projects.
- Often, resources are devoted to development activities only after negative customer feedback or when profits have decreased indicating that there are problems (therefore, innovation activities are often reactive, instead of being proactive).
- Personnel training has an important role: either it takes place in a form of courses provided outside a company and then distributed to others in an organization or in a form of mentoring between a senior and a young member of a company. Only then is personal knowledge diffused in the organization so that it gradually has an impact on overall performance and services.
- However, most of the KIBS companies do not systematically manage their knowledge portfolios, and there are only few companies which have intentionally made efforts to diffuse the accumulated stock of knowledge, now often embodied in few key experts, more evenly within their organizations.

After this short introduction I discuss the spatial dimension of innovation activities and introduce the concept of local innovation environment. Thereafter I present some notions on the innovation-related, inter-organizational networks and contacts. I briefly examine the above-mentioned themes also empirically from the point of view of the new media and software companies and pose the following research questions:

1. Which are the key features of the new media and software companies' innovation activities?
2. What kinds of innovation networks do these companies have?
3. What kinds of inter-personal networks do their executives maintain?
4. What is the role of geographical proximity in their innovation activities?

Spatial Dimensions of Innovation Activities

The Basic Concepts and Ideas

The relationship between companies' innovation activities and their local operational environment is an interesting research theme, with substantial societal relevance. Consequently many alternative theoretical approaches and concepts have been developed, including (regional) clusters (e.g. Porter 1990; 1998), new industrial districts (e.g. Harrison 1992; Cossentino et al. 1996) innovative milieus (e.g. Camagni 1991; 1995) and (regional) systems of innovation (e.g. Braczyk et al. 1998; Howells 1999). All of these concepts are based in one way or another on the idea that certain kinds of activities have clustered spatially in certain territories. All recognize that the ways in which the local operational environment can promote innovation are very intricate, but should not be exaggerated. Many knowledge-intensive companies are very loosely or not at all connected to their local environments, making bold generalizations related to the role of regions debatable (cf. Miettinen et al. 1999; Miettinen 2002).

A local operational environment boosts companies' innovation activities only when the relationships among companies (collaborative and competitive) and between companies and other organizations are characterized by intensive innovation or mutual learning orientation. In that case, we can use the concept of innovative environment or milieu, which can simply be defined as an existing set of companies, institutions and social networks which creates a potential base for the emergence of localized innovation networks (cf. Cooke 1998, 2002; Camagni 1991). All companies do not have the same kind of relationship with the local operational environment, because each company follows its own strategy in this respect as well. Yet some general behaviour patterns can be found. For example, on the basis of their empirical study, Kautonen et al. (2002) argue that the companies referred to as KIBS companies and traditional, supplier-dominated manufacturers are tightly integrated into their regional environment by having locally or regionally located partners in their innovation networks. These partners are usually either key customers, suppliers, partner companies or other organizations such as public business service providers. These types of companies are usually also the smallest ones. Furthermore Kautonen et al. argue that KIBS companies have an important role in regional innovation environments, because they mostly have very close network-type relationships with their innovation partners and these are often located in the same region (cf. e.g. Muller and Zenker 2001).

Three Levels of the Local Innovation Environment

When analysing a certain region, locality or territorial agglomeration, it might be useful to prefer the concept of innovation environment to the concept of innovative environment or milieu, because the former does not have the same normative flavour as the latter. Namely, the local innovation environment can be either good or bad or it may make no difference at all. I preliminarily argue that these factors and elements should be analysed at three different levels: *1. the structural and institutional level, 2. the level of organizational relationships* and *3. the level of individuals.* Each level has certain characteristics and dynamics that are necessary to make a local innovation environment (e.g. an industrial agglomeration) innovative. I discuss each three levels here, only briefly.

The structural and institutional level. The basic business and institutional structure of the agglomeration has a significant influence on the dynamics of an agglomeration. This holds true also from the point of view of innovation activities. Therefore, when analysing a local innovation environment, attention should be drawn at least to the following factors and elements: the number and nature of companies and business units, educational institutions, science and technology base (e.g. universities, research institutes and private RandD units), specialized private, semi-public and public business services (e.g. financing, consultancy, technology transfer and incubation services) and interest groups (e.g. trade and entrepreneurial associations, chambers of commerce), local authorities. When considering the institutional setting, the concepts of local institutional density and institutional thickness become very interesting. These concepts refer to the local presence of numerous different institutions collaborating synergetically to attain a somewhat common goal, guided by partly shared norms, values and understanding (cf. Amin and Thrift 1994). For example, Kolehmainen et al. (2003) argue that the educational and research institutions can have a very crucial role in the creation and further development of a good local innovation environment, especially in knowledge-intensive "technopoleis" (cf. Gibson and Stiles 2000).

The level of organizational relationships. A city or an agglomeration does not innovate by itself, but it can support the innovation activities of organizations. The structural and institutional setting of a local innovation environment forms one possible basis for these activities which are increasingly inter-organizational and network-based in nature. Consequently, when analysing the local innovation environment from the point of view of companies, attention should be drawn to the presence of demanding customers, advanced suppliers and subcontractors, technology and other partner companies and universities and research institutions, for example. Naturally the local presence of these kinds of organizations is not enough, because their nature (e.g. the level of know-how, ability to co-operate and resources) determines whether or not there are possibilities for mutually synergetic co-operation. In addition to the co-operative local inter-organizational relationships, the local competition between different organizations, mainly between companies, can stimulate the innovation activities.

The level of individuals. Inter-organizational relationships can to some extent be reduced to relationships among individuals working in organizations. This notion sets up the argument that the role of individual people is very remarkable in local innovation environments. The social nature of inter-organizational relationships is only one dimension of the role of individuals, because skilled workers and experts usually have extensive, work-related personal networks which facilitate the seeking of rare, reliable, or in other ways valuable information and knowledge. From the spatial point of view, it can be argued that geographical proximity does matter in the formation and utilization of these individual contacts despite the advanced information, communication and transportation facilities. However, from the point of view of individuals, the local innovation environment cannot be reduced to only a 'platform' for localized social networks and relationships. There are also other dimensions. For example, for the individuals, the local innovation environment should also be 'a creative problem-solving environment', which is a concept that refers to the presence of diverse and high-quality career and further education opportunities (cf. Raunio 2001), for example.

The Local Innovation Environment and Inter-organizational Networks

Thus, the local innovation environment also consists of a set of different kinds of inter-organizational relationships and contacts. Geographical proximity facilitates the birth and utilizations of these relationships which can be either collaborative or competitive in nature. The inter-organizational collaboration and competition increasingly take place even at the same time. This notion produced a new concept of "co-opetition" (see e.g. Brandenburger and Nalebuff 1996). The versatile interaction among companies and other organizations takes usually place in networks and when the inter-organizational interaction is related to the innovation activities a concept of innovation network can be used. The local innovation environment's level of organizational relationships refers to a great extent to the localization or the local dimension of the innovation networks. However, these innovation-related networks or other contacts are not necessarily (and should not be) localized, because companies seek, or at least they should seek, the most suitable partners in co-operation. Still, it can be argued that geographical proximity may facilitate the interaction in these networks and relationships. All networks are not alike and subsequently I present some notions concerning the inter-organizational and especially inter-firm networks.

Concerning inter-firm networks, one basic classification is to divide firms into two groups, of which the former often require formal co-operative or joint business arrangements while the latter usually remain informal (Rosenfeld 1996, 248):

- 'Hard' networks: three or more companies joining together to co-produce, co-market, co-purchase, or co-operate in product or market development
- 'Soft' networks: three or more companies joining together to solve common problems, share information or acquire new skills.

Hard networks often require formal co-operative or joint business arrangements, while soft networks usually remain informal. This typology can be augmented with a typology of business networks introduced by Lechner and Dowling (2003). They have studied empirically networks of high-growth entrepreneurial companies in the IT industry and explored how these companies grow through the use of external relations and become competitive. This typology and key features of the networks can be summed up as follows:

- *Social networks*: Social networks are used to create the first business networks of a start-up company. These networks are not yet real inter-organizational networks, but they can develop into useful vertical and horizontal ties. Social networks automatically lead to trust-based business relationships that would otherwise be difficult to build. Although these network-like relationships are a valuable asset for a newly established company, they have only a limited impact on subsequent firm development.

- *Reputational networks*: Start-up companies seek new and highly visible partners that provide a certain kind of reputational promotion for them in addition to the real substance of collaboration. However, this kind of partnership or network relationships are not trouble-free. Usually the reputable partners require something very valuable in exchange and the "terms of trade" are not necessarily favourable for the start-up (e.g. fear of loosing the own technology). Reputational networks are still very useful. They first signal to the market that the new company is a viable partner and help start-ups to overcome liability of newness. This facilitates the further network building. The lack of reputational networks may even constitute a growth barrier for a start-up.
- *Co-opetition networks*: In today's business environment co-operation with competitors ("co-opetition") is not uncommon. Especially the entrepreneurial companies seek actively relationships with competitors. Co-opetition networks are an important source of entrepreneurial companies' flexibility and growth and allows them to focus on core competencies. Namely, co-opetition networks have two basic functions: Firstly, a company can hand over or subcontract a project to another company and in that way avoid the loss a client because of limited company size and capacity. Secondly, companies rely on certain partners in order to be able to take care of larger projects (e.g. total system solutions). This leads to collaboration with companies that are usually at least indirect competitors. Co-opetition networks are also trust-based and they require a high degree of reciprocity. They are also usually regionally embedded, because finding a similar and suitable competitor in terms of size and attitude towards co-operation is easier in the home region for four reasons: (1) regional culture influences corporate culture; (2) finding a partner is easier if the company has a functioning information network which is usually mainly regional; (3) a successful agglomeration of companies attracts more business increasing the probability that the favour will be returned; and (4) trust is built over time and through frequent and intensive interaction (e.g. face-to-face contacts). Given time and energy constraints, geographical proximity matters.
- *Marketing networks*: Marketing networks are relationships with other companies that enable the central company gain better market information, reach new markets or gain new clients. The forms of partnership might range from distribution alliances to real client procurement. Marketing networks are not detached from other business relations and they overlap frequently with social and co-opetition networks, for example. Customers and suppliers may play an important role in entering new markets and penetrating the existing markets better. It seems to be so that companies' generally extensive networking behaviour also contributes to the formation of efficient marketing networks.
- *Knowledge, innovation and technology (KIT) networks*: KIT networks are relationships with other companies that allow the access to and the

creation of new (technological) knowledge and innovation. Both vertical and horizontal ties are possible and the main feature of these relationships is the interactive knowledge-sharing with people who trust each other. It can be claimed that companies are increasingly building partnerships and engaging in pure learning alliances that facilitate also the creation of new knowledge. Innovation processes can be initiated spontaneously, randomly and/or systematically. There are also a variety of modes by which companies develop innovations due to the relations with other companies. The creation of (regionally embedded) knowledge-sharing and learning spaces enables trust-based networking.

It can be argued that from the "recursive" innovation model's point of view all of these network types can also be a part of company's innovation networks. Lechner and Dowling (2003) note also that the various networks overlap and that certain networks require other networks in order to enable relations. For example, reputation and co-opetition networks often require social or KIT networks. From the local innovation environment's point of view, it is notable that some networks, such as social and co-opetition networks, are localized more strongly than the others because of the internal dynamics within these networks. However, some networks, such as knowledge, innovation and technology networks, can be both local and global. The good local innovation environment may facilitate the birth and development of these kinds of networks.

Local Innovation Environment and Inter-personal Networks

Innovation is a process in which various pieces of knowledge are put together in order to achieve something new. Naturally, the rate of newness varies from innovation to innovation. Companies acquire the knowledge needed in innovation from several sources. Some pieces of knowledge are explicit, such as patents, licences, manuals or technical blueprints. Some pieces of knowledge are embedded in concrete artefacts, such as machinery and new kinds of raw materials and intermediate products. The intangible nature of services is reflected also in this issue. It can be argued, that the embodied, embrained or encultured knowledge plays a big role in the innovation activities of knowledge-intensive business service companies (cf. Blackler 1995). These types of knowledge are usually anchored to the professionals working in these companies. This notion leads us to the issue of the knowledge sources and channels of these professionals. The knowledge sources and channels can crudely be divided into two categories: explicit and social. The explicit knowledge sources on the individual level are greatly the same as on the company level, namely instruction manuals, blueprints etc. In addition to these, professional acquire knowledge from trade magazines, Internet, fairs, television, just to name few. The concept of social knowledge sources and channels refers to those people who provide useful information and knowledge to a given professional.

Attention has been paid to the inter-personal linkages and social networks also from the point of view of information for a long time. The article by Mark Granovetter (1973) titled "The strength of weak ties" is a classic in this field. One of the main points of that article is that weak ties, i.e. ties between persons who do not know each other so well, are important as channels of new and unexpected information. Broadly speaking, this is the strength of the weak ties.

Some of the Granovetter's (1973) ideas can be applied in the development of the concept of local innovation environment on the individual level. The inter-organizational innovation networks are usually confirmed by some kind of juridical arrangements, for example by non-disclosure agreements. The building of these networks requires usually also economic investments. These features make them strong and explicit. From the individual point of view these inter-organizational networks are usually a basis for the birth of strong ties between persons who are responsible for the inter-organizational link on the behalf of their own organization. However, professionals have strong ties also irrespective of their organization's networks and contacts. These ties are usually born through the current or previous job or through joint studentship. It can be argued that these strong ties are still very important information and knowledge channels although they may not offer very new and unexpected pieces of knowledge. Instead, they may provide "filtered", insightful and reliable knowledge and support for decision-making and problem-solving. These social networks have naturally also many other dimensions than serving only as information and knowledge sources and channels. Usually it is about friendship also outside the professional context.

From the "granovetterian" point of view, these strong ties or linkages are not enough, but they have to be complemented by weak ties. Attention has been paid to these weak ties or linkages also within the territorial innovation model (TIM) debate (see e.g. Camagni 1991). In this context weak ties refer usually to open, unofficial and implicit linkages between organizations that are local in nature. These linkages are also strongly inter-personal and social and they are not usually been built consciously, but they have emerged through various different kinds of occasions (e.g. seminars, conferences, meetings). Although the weak ties are not target-oriented, they may be very useful as far as the knowledge acquiring is concerned. They may channel information and knowledge about the current developments within a certain branch, technological trends, competitors, customer needs etc. The information is usually non-specific, fuzzy, splintered and it may even not be very reliable. However, these pieces of information and knowledge may provide some seeds for new ideas and insights. Therefore the weak ties and the "local buzz" generated by them are of considerable value also from the innovation activities' point of view. (see Kolehmainen 2004) In addition, this is especially the case within knowledge-intensive business services. Additionally, it has to be emphasized that weak ties described above are not an "exclusive right" of local innovation environment, but they can be regional, national or international. Still, it can be argued that the geographical proximity facilitates and enhances the birth of the weak ties, because they are based on multidimensional face-to-face interaction that enables the exchange of complex implicit knowledge (Storper and Venables 2002; 2003).

Some Empirical Findings from New Media and Software Companies

Survey

The empirical part of this chapter is based on a survey that was aimed at 848 Finnish new media and software companies (business units). The survey was conducted via the Internet by using a web-based questionnaire. The questionnaire was designed to be fairly extensive, but easy to complete. The managing directors or directors of business units were approached by e-mail twice and altogether 105 answers were garnered. Accordingly, the response rate was 12.4 per cent, which is very low. Using the Internet can be justified by the type of the target companies: the Internet is a natural channel to approach software and new media companies. On the other hand, the managing directors and directors of business units of those companies receive typically a massive number of emails and responding to surveys is not by any means in their top priority. However, the number of answers as such is acceptable. In general, the response rate in web-based surveys tends to be lower than in surveys in which the data collection is conducted by using different technologies (e.g. mail or phone). In any case, both lower and higher response rates have been achieved in research settings somewhat similar to this one (see e.g. Kaufmann et al. 2003; Varis et al. 2004). The sample is otherwise fairly representative.

Basic Description of Survey Respondents

The survey data consists of 105 responses. A vast majority (85 per cent) of them came from independent companies, but there were also some answers from companies belonging to a bigger company group. About 68 per cent of the companies are located only in one place and the rest have two or more business units. The average age of the companies is eight years and a fifth of companies are two years old or younger and a third of the companies have existed for more than ten years. The software service companies formed the oldest group, whose average age is almost ten years. The youngest company group was formed by companies that classified themselves into the group "other companies". The average age of the new media and software companies varies between seven and eight years.

The biggest proportion of the companies (37.5 per cent) is located in the capital region. Among other locations were Tampere (17.3 per cent), Jyväskylä (9.6 per cent) and Turku (5.8 per cent). Oulu, the "North Technopolis", is not presented very well in the survey data. About 18 per cent of the companies are located in "operational micro environments" which are directed especially for that kind of companies (e.g. technology centres or business incubators). In the year 2001 the

Table 13.1   Companies' products and services

Product / service	Number	%
Customer-specific or tailored software	62	59.0
Software products (business-to-business market)	61	58.1
Design and implementation of WWW services	49	46.7
Embedded software, software services, software subcontracting	27	25.7
Providing WWW services and contents	27	25.7
Design and implementation of multimedia solutions and products	17	16.2
Software products (customer market)	15	14.3
Internet-based advertising and marketing	9	8.6
Connection and hosting services	7	6.7
Knowledge-intensive services related to all of the above as part of company's other activities	75	71.4

average number of employees in the respondent companies was 13.6 persons and the average turnover of employees seven per cent. The average age of personnel was about 34 years and the average share of employees with an academic degree was nearly 50 per cent. Also the financial position of the companies was examined by asking the companies' turnover and operating income during the three latest years (1999–2001). The average turnover of all the companies was 0.86 m€ in the year 1999 and in the year 2001 it was 1.38 m€. The turnover increased in all company groups, namely the new media companies, software product companies, software service companies and "other companies". The software service companies had the biggest operation income.

The respondent companies seem to have a quite wide product and service range (Table 13.1). More than a half of the companies produce customer-specific or tailored software (59.0 per cent) and software products to business-to-business markets (58.1 per cent). Approximately 47 per cent of companies design and implement WWW services for their customers, which is typical of the new media companies. A vast majority (71 per cent) of the companies offer knowledge-intensive services (e.g. training and consultancy) related to the company's other products and services. It seems to be typical of new media and software companies to be active in many product and service categories. The respondent companies have on average 3.3 product or service categories. Approximately 36 per cent of the respondent companies categorize themselves as software companies, 27 per cent as software product companies and 21 per cent as new media companies. The rest of the companies consider themselves to belong in the group "Other companies", although they produce some kind of software or new media product or services.

Companies' Innovation Activities and Networks

Targets of Innovation Activities

The examination of the new media and software companies' innovation activities clearly reveals the importance of the further development of existing products and services (Table 13.2). Almost 31 per cent of the companies consider that it is the most important target of innovation. Many companies also pay a lot of attention to the development of new products and services, which is related to the modification of product or service portfolio. One fifth of the companies chose that as the most important target of innovation. Approximately 14 per cent of the companies see that the development of new business ideas and concepts is the most important target of innovation. One tenth of companies put the greatest effort on the development of collaborative relationships with other companies and organizations. It seems to be that there are no major differences between different company groups.

The matter of the most important target of innovation is also a matter of time-span. In the short run, the further development of current products and services may be very important but, in the long run, the development of new products and services, new business ideas and concepts or new technology may be even more important and especially more strategic. Still, it is quite remarkable that the other possible targets of innovation activities were not considered to be very important, although they may be a big part of the development of the whole business. This notion alludes not only to the product or service centredness of the new media and software business, but also to the fact that the general understanding of innovation has still not changed. However, the emerging role of collaborative relationships with other companies and other organizations as a target of innovation is highly compatible with the new, "recursive" innovation model.

Innovation Networks and "Linkage Groups" in Innovation

As highlighted in the theoretical discussion, the companies do not innovate in "vacuums", but as parts of versatile, interactive innovation networks. These

Table 13.2   Targets of innovation activities

Target of innovation activities	Number	%
The further development of existing products and services	30	30.6
The development of new products and services, the modification of product / service portfolio	20	20.4
The development of new business ideas and concepts	14	14.3
The development of collaborative relationships with other companies and organizations	10	10.2
The development of totally new knowledge or technology	8	8.2
The searching of new market areas and customers	7	7.1
The development of new distribution channels	7	7.1
The development of company's core processes (e.g. the streamlining of R&D projects or customer projects)	1	1.0
The development of product and company brand(s)	1	1.0
Total	**98**	**100.0**

explicit, inter-organizational networks are augmented by implicit, inter-personal networks. These social networks are an important part of the individual level of the local innovation environment. According to the survey, many organizations can be innovation partners for the new media and software companies (Table 13.3). However, the most important partnership relationships related to innovation activities seem to be quite concentrated in few linkage groups. Companies have on average approximately three (2.71) external linkage groups which are considered to be important or very important in innovation activities. In this sense, there are no big differences between company groups: the lowest average figure was in the group of software service companies (2.27) and the highest in the group "other companies" (3.06).

Table 13.3   The most important partners in innovation networks / 'linkage groups' in innovation

"Linkage Group"	The most important		2nd most important	
	Number	%	Number	%
Company's other business units / other companies in the same consolidated corporation	6	5.7	12	11.9
Customers	**84**	**80.0**	15	14.9
Competitors	1	1.0	7	6.9
Suppliers of software development tools and environments	2	1.9	**18**	**17.8**
Suppliers of software components and platforms	3	2.9	6	5.9
Software service suppliers			2	2.0
Content service suppliers			3	3.0
Consultancy companies / other non-technological KIBS providers	1	1.0	3	3.0
Private financiers	2	1.9	8	7.9
Universities / research institutes	1	1.0	12	11.9
Polytechnics and other educational institutes	1	1.0		
Public, national support organizations (e.g. Tekes)	1	1.0	5	5.0
Local support organizations			2	2.0
Other	3	2.9	8	7.9
Total	**105**	**100.0**	**101**	**100.0**

According to the survey, customer companies are absolutely the most important linkage group or network partner in new media and software companies' innovation activities. Exactly 80 per cent of the companies named their customers as their most important innovation partners, which is by no means a surprise, given the targets of innovation activities in particular: It is very natural to develop the current and totally new products and services in close collaboration with the customer. That is the case particularly in B2B business. When asking the most important external innovation partner, the answer alternative "Company's other business units / other companies in the same consolidated corporation" received the second biggest frequency. Because these two responses were expected, also the second most important innovation partner or linkage group was asked, and this question divided the respondent companies more evenly. It also highlighted the importance of technology and knowledge suppliers in the innovation activities of new media and software companies. Namely, suppliers of software development tools and environments were most frequently chosen to be the second most important innovation partner or linkage group. Almost one fifth of respondent companies had this view. On the other hand, this question brought the role of universities and research institutions to the fore. Approximately 12 per cent of the companies named them as the second most important partner and the figure is exactly the same as for the companies' other business units and other companies in the same consolidated corporation.

The Spatial Dimension of Innovation Networks

The spatial dimension of the companies' explicit, inter-organizational innovation networks was examined by asking the importance of collaboration with the organizations belonging to the most and second most important innovation partner group locating in different regions (Table 13.4). The basic idea was to chart the location of the core of companies' innovation networks. Generally speaking, it can be claimed that a locality or region is a natural "platform" for the formation of innovation networks. Approximately 30 per cent of companies see that collaboration with local or regional organizations belonging to the most important innovation partner group is important or very important. However, the national scale (excluding the company's own location) has in this sense the biggest proportion (38.4 per cent). One fifth of the companies consider the collaboration with organizations within the European Union[1] important or very important. A corresponding question, considering the collaboration with organizations belonging to the second most important innovation partner or linkage group, results in a quite similar distribution. However, the role of the European Union seems to be slightly smaller and correspondingly the role of North America more considerable. Still, these results do not allow us to draw very profound conclusions.

There seems to be some differences in the spatial dimension of innovation networks between different company groups. For instance, the innovation networks

---

1   Survey was conducted before the latest enlargement of EU in May 2004.

Table 13.4   Spatial dimension of innovation networks

Location	The most important partner group	2nd most important partner group
The same technology centre, science park etc., %	7.7	8.6
The same city / town, %	28.9	*23.1*
The same region, %	*29.8*	21.2
Other regions in Finland, %	**38.4**	**28.8**
Scandinavia or the Baltic countries, %	10.6	5.7
Other EU countries, %	20.2	12.5
North America, %	10.6	13.5
Other countries, %	5.8	5.7
N	104.0	104.0

of the new media companies are more frequently local or regional. Half of the new media companies consider the collaboration with organizations belonging to the most important innovation partner or linkage group locating in the same region as important or very important and the figure is almost as high in the case of the same city or town. Also the networks of software service companies seem to have a fairly strong local and dimensional emphasis. Correspondingly, the software product companies have more clearly twofold innovation networks: they see both regional (and local) and international contacts as almost equally important.

Executives' Inter-personal Networks

The Role of Inter-personal Networks

As suggested above, individual professionals' personal linkages and social networks play a certain role in the companies' innovation activities. This role can be considerable and it is related to the acquiring of information and knowledge needed in innovation processes. Especially newly established and other small companies are dependent on their key professionals' personal linkages and social networks. It can be even so that these social networks and companies explicit inter-organizational innovation networks are overlapping to great extent. In this chapter some empirical views on these social networks are presented. They are based on the same empirical data as previously presented results concerning the companies' innovation activities and

Table 13.5   Areas of innovation-related information

Areas of innovation-related information	Number	%
Development of new products / services	39	37.5
Technological changes in the long run	20	19.2
Changes in markets and customer needs in the long run	17	16.3
Day-to-day management	8	7.7
Supplier and partner companies	8	7.7
Competitors (competitive intelligence)	5	4.8
Changes in markets in the short run	4	3.8
Financing	2	1.9
Activities of research and educational institutes and support organizations	1	1.0
Total	**104**	**100.0**

networks. These results reflect the managing directors' and business unit directors' (later executives') personal linkages and social networks. These networks exemplify the strong ties between people. Unfortunately, the role of weak ties is not empirically dealt with in this article.

According to the survey, the information and the knowledge the strong ties, i.e. the close inter-personal networks provide to the executives are related to many issues (Table 13.5). However, most of the executives' personal linkages seem to be related to the development of new products and/or services. Inter-personal networks have also importance in the monitoring and evaluation of technological changes and changes in markets and customer needs in the long run. All these issues are naturally intertwined because the development of new products and services in software and new media companies – as in most companies – is dependent on the one hand on the new technological possibilities and on the other hand on the emerging customer needs and new ways to fulfil those needs. The social linkages are used also as means for the watching and selection of supplier and partner companies and they are also of some value in respect of competitive intelligence. In addition to these issues, inter-personal networks provide support for day-to-day management.

Table 13.6   The background organization of key contacts

Background organization	%
Company's other business units / other companies in the same consolidated corporation	22.1
Customers	**76.0**
Suppliers and partner companies	*55.8*
Competitors	23.1
Consultancy companies' / other (non-technological) KIBS providers	29.8
Other companies	25.0
Private financiers	14.4
Universities / research institutes	24.0
Polytechnics and other educational institutes	11.5
Public, national support organizations (e.g. Tekes)	10.6
Local support organizations	10.6
Branch specific or business association	9.6
Other	5.8
N	104.0

Orientation of Inter-personal Networks

Executives use their inter-personal linkages as information or knowledge channels in issues that are closely linked to their companies' concrete innovation activities. Therefore, it is only natural that executives' key contacts are often in the same organizations with which their companies collaborate within innovation processes (Table 13.6). Thus, the role of customers and supplier and partner companies is essential. According to the survey, executives have important key contacts on average in three (3.18) linkage groups. Altogether three out of four (76 per cent) executives have personal contacts with customers and 56 per cent of them have personal contacts with people representing supplier and partner companies. Approximately every third

Table 13.7   The relationships' birth modes

Relationships' birth mode	%
Current job	**83.7**
Previous job	*48.1*
Exhibition, conferences, seminars and other events related to the business	34.6
Associations or interest groups related to the business	10.6
Studentship	25.0
Educational events, further education and training	19.2
Joint hobbies	14.4
Friendship, acquaintanceship or kinship	29.8
Belonging to common religious, linguistic and other groups	3.8
Other birth mode	9.6
N	104.0

of executives have personal contact with people in consultancy companies and in other (non-technological) KIBS providers which are not customers, suppliers or partners of their company. Respectively, roughly every fourth of executives have important contacts with people in companies in different branch, in universities and research institutes or even in competitor companies. The role of public support organizations from this inter-personal networks' point of view seems to be quite marginal.

Inter-personal Networks' Birth

The respondents were asked how their inter-personal networks have come into being (Table 13.7). According to the survey, executives' important inter-personal networks are tightly connected to their current job. A vast majority of the executives (83.7 per cent) stated that important social linkages have come into being through their current job. Additionally, roughly a half of the respondents (48.1 per cent) stated that the previous job has been an important way to create social linkages of great value.

Table 13.8   Spatial dimension of inter-personal networks

Location	%
The same technology centre, science park, etc	10.6
The same city / town	**54.8**
The same region	33.7
Other regions in Finland	*53.4*
Scandinavia or the Baltic countries	11.5
Other EU countries	22.1
North America	15.4
Other countries	5.8
N	104.0

Exhibitions, conferences, seminars and other events related to the business seem to have surprisingly considerable role as inter-personal networks' birth modes, because more than every third of respondents mention that.

Friendship, acquaintanceship or kinship and joint studentship seem also to have some importance as source of useful social linkages also in respect of business and innovation. Interestingly enough, among newly established companies (over two years) the role of last-mentioned relationships' birth mode is more considerable than among older companies. In addition to studentship, different kinds of educational events, including further education and training, offer opportunities to establish useful relationships. Also the joint hobbies serve this same purpose to some extent, especially for the executives of newly established companies. These findings are very coherent to the notions by Lechner and Dowling (2003) concerning the role of the social networks in the development of newly established companies. According to the survey, the other possible relationships' birth modes seem not to have significant role at all.

The Spatial Dimension of Inter-personal Networks

Companies' explicit innovation networks are most often national and increasingly international. In comparison it can conjectured that the inter-personal network are more strongly localized because presumably the importance of face-to-face interaction is bigger within social networks than in explicit inter-organizational innovation networks. The survey provides some support for this hypothesis (Table 13.8). More than half of the respondents (54.8 per cent) have important personal contact persons in the same city or town in which they work themselves. Still, almost as many of the respondents have this kind of social context somewhere else in Finland than their own city or region.

The "micro innovation environments", such as technology centres or science parks seem to have subtle importance as platforms for the formation of inter-personal networks. Approximately eleven per cent of the respondents have an important contact person within that kind of environment. Thus, technology centres and science parks seems to facilitate the networking of people to some extent. Despite the local bias, inter-personal networks also have an international dimension. More than a fifth of the respondents have important social contacts in other EU countries (excluding those Nordic countries that belong to the EU). The corresponding rate concerning North America is 15.4 per cent and Nordic and Baltic countries 11.5 per cent. The EU figure in particular is quite high.

Conclusions and Discussion

The concept of innovation has changed and the new thinking accentuates especially the importance of companies' external interactive relationships in innovation activities. On the other hand, the role of continuous learning and incremental innovations is highlighted. At the same time, increasing attention has been paid to the innovation in the service sector and pre-eminently the knowledge-intensive business services (KIBS) have been in the focal point of research. The innovation activities in the KIBS companies differ to some extent from the innovation in manufacturing companies. For example, product development is integrated with customer projects and usually carried out side by side along with normal daily routines. In addition to the innovation activities of their own, the KIBS companies may have many important roles in their customer companies' innovation activities. The KIBS companies are usually quite strongly integrated in their local or regional operational environment.

These notions together have partly brought up again the discussion of the importance of the local operational environment in companies' innovation activities. Geographical proximity is one of the profound elements of the local operational environments. However, the ways in which they can promote innovation are very complex and intricate. To bring some clarity to this issue, the concept of local innovation environment was introduced. It was argued that a local innovation environment consists of many interrelated elements ranging from the institutional setting to the behaviour of individuals. It was also argued that the local innovation environment can be divided into three levels: (1) the structural and institutional level, (2) the level of organizational relationships and (3) the level of individuals. At each level certain characteristics are necessary for the local innovation environment to affect companies' innovation substantially.

Many of these ideas were examined empirically from the experiences of Finnish new media and software companies. It can be claimed that the basic features of innovation activities in the new media and software companies are quite similar to those of most of the KIBS companies. This is the case at least from Finnish perspective. The innovation activities are incremental in nature; most of the companies see that the most important area of their innovation activities is to develop existing product portfolios by developing new products. This product development can described to some extent incremental, because the products are in many cases based on previous products and totally new knowledge or technology is applied quite rarely. However, it can be argued that these survey results reflect the executives' understanding of innovation in general. For example, the informal and incremental development of internal processes and procedures may not be considered as an innovation, although they may have a major impact on the profitability of the business.

Analysis of the inter-organizational innovation networks of the new media and software companies revealed that these are not separate from their basic, operational business networks. A characteristic feature of KIBS companies is that their key customers are their most important external innovation partners in incremental product innovations. In many cases companies are trying to create very close, collaborative relationships with their customers. The customer-driven strategy in innovation activities is in many cases very advisable, if it does not lead to a "lock-in", to a situation in which tight relationships hinder the development and growth of the company. From the technological point of view, it seems that the most important partners in new media and software companies' innovation activities are the commercial technology providers, such as suppliers of software development tools and environments as well as software components and platforms. For example, the role of university-company collaboration does not seem to be very crucial in new media and software businesses. It seems to be so that the technology *per se* is not a problem, it is all about innovative appliance of known technologies in an innovative way that meets and even exceeds customer requirements.

The analysis of executives' inter-personal networks from the innovation activities' point of view revealed that executives use their inter-personal linkages as information or knowledge channels on issues that are closely linked to their companies' concrete innovation activities, such as development of new products and services. Therefore, executives' important key contacts' background organizations represent usually the same organization groups with which their companies collaborate. This notion underlines the certain overlap between inter-organizational innovation networks and executives' (and more broadly, professionals') inter-personal networks. This same overlap is also reflected by the fact that a majority of executives' important social linkages have become into being through their current job. On the other hand, the survey indicated for the executives of newly established companies, the other means and forums of building inter-personal networks are of notable importance.

The innovation networks of the companies seem to be localized to some extent. About one third of the companies have very important innovation partners in the same city or in the same region, although the national level seems to be the most important level of inter-organizational innovation network formation. However, based on the previous studies on KIBS companies, it could have been expected that the innovation networks would have been even more localized. The spatial scope of the companies' innovation networks seems to be both local and national. This is in line with the notion that the Finnish innovation environment is usually characterized to be locally and nationally intertwined, underlining the small size of the whole country. On the one hand, in most regions there are not enough companies to form 'critical mass' from the innovation activities' point of view. On the other hand, in a small country it is very easy to build collaborative networks nationwide within a certain industry, as suitable partners are quite easy to recognize and approach. In addition to these features specific to Finland, one could argue that companies have always – irrespective of their location – to find the right balance of localized sources of interaction and those of national and international ones to maintain and develop the competitiveness and to avoid problems, such as lock-in situations (cf. Nachum and Keeble 2003).

The spatial dimension of executives' inter-personal networks is very similar to the spatial dimension of inter-organizational innovation networks. This is natural because these networks are overlapping in nature. However, there are certain indications in the data that inter-personal networks are more strongly localized than inter-organizational innovation networks. This tendency seems not to be very strong, but it is an expected one in the light of theory: the so-called "local buzz" and the weak ties related to it can turn into strong inter-personal ties that may be useful from the innovation activities' point of view. Still, very strong conclusions from these results can not be drawn. From the inter-personal networks' point of view, the national level is still very important, largely due to the same reasons as in the case of inter-organizational networks. Additionally, it should be noted that the empirical results presented in this article did not deal with the weak ties and their spatial and other features at all. That is a matter of further research that requires also totally different kind of methodology.

All the results together support the assumption that the local innovation environment and its key element – geographical proximity – do matter in innovation, but it should not be exaggerated. Still, from the proximity point of view, other forms of proximity need to be taken into account as well. Different forms of proximity, such as geographical, social, cognitive, organizational and institutional proximity, play some role in the formation of inter-organizational networks and collaboration related to innovation. As Boschma (2005) argues, too much or too little of each form of proximity is detrimental to inter-organizational learning and innovation. Additionally, the different forms of proximity are usually intertwined in many ways and they may reinforce or substitute each other depending on the case. Every company faces these fundamental questions when building its collaborative networks related to innovation and the choices of companies results in the structure and dynamics of the local innovation environment. To capture theoretically or not to mention empirically all the features and mechanisms of the local innovation environment having influence on the companies' innovation activities and performance is extremely difficult. This reflects nicely the complexity of the reality as well.

References

Amin, A. and Thrift, N. (1994), 'Living in the Global', in Amin, A. and Thrift, N. (eds.) *Globalization, Institutions, and Regional Development in Europe* (Oxford: Oxford University Press).

Blackler, F. (1995), 'Knowledge, Knowledge Work and Organizations: An Overview and Interpretation', *Organization Studies* 16: 6, 1021–1046.

Boschma, R.A. (2005), 'Proximity and Innovation: A Critical Assessment', *Regional Studies* 39: 1, 61–74.

Braczyk, H-J., Cooke, P. and Heidenreich, M. (eds) (1998), *Regional Innovation Systems: The Role of Governance in Globalized World* (London: UCL Press).

Brandenburger, A.M. and Nalebuff, B.J. (1996), *Co-opetition* (New York: Currency-Doubleday).

Camagni, R. (ed.) (1991), *Innovation Networks: Spatial Perspectives* (London and New York: Belhaven Press).

Camagni, R. (1995), 'The Concept of Innovative Milieu and its Relevance for Public Policies in European Lagging Regions', *Papers in Regional Science* 74: 4, 317–340.

Cooke, P. (1998), 'Introduction. Origins of the Concept', in Braczyk, H.-J., Cooke, P. and Heidenreich, M. (eds), *Regional Innovation Systems* (London: UCL Press).

Cooke, P. (2002), *Knowledge Economies: Clusters, Learning and Co-operative Advantage.* (London and New York: Routledge).

Cossentino, F., Pyke, F. and Sengenberger, W. (eds) (1996), *Local and Regional Response to Global Pressure: The Case of Italy and its Industrial Districts* (Geneva: International Institute for Labor Studies, Research Series 103).

Faulkner W. (1995), 'Getting Behind Industry-Public Sector Research Linkage: A Novel Research Design', *Science and Public Policy* 22: 5, 282–294.

Gibson, D.V. and Stiles, C.E. (2000), 'Technopoleis, Technology Transfer, and Globally Networked Entrepreneurship', in Conceicao, P., Gibson, D., Heitor, M. and Shariq, S. (eds), *Science, Technology, and Innovation Policy: Opportunities and Challenges for the Knowledge Economy* (London: Quorum Books).

Granovetter, M. (1973), 'The Strength of Weak Ties', *American Journal of Sociology* 78, 1360–1380.

Harrison, B. (1992), 'Industrial Districts: Old Wine in New Bottles?', *Regional Studies* 26, 469–483.

Kaufmann, A., Lehner, P. and Tödtling, F. (2003), 'Effects of the Internet on the Spatial Structure of Innovation Networks', *Information Economics and Policy* 15: 3, 402–424.

Kautonen, M. (2001), 'Knowledge-Intensive Business Services as Constituents of Regional Innovation Systems: Case Tampere Central Region', in Toivonen, M. (ed.) *Future Prospects of Knowledge-Intensive Business Services* (Helsinki: Publications of the Helsinki Region Centre for Business and Employment Development).

Kautonen, M., Kolehmainen, J. and Koski, P. (2002), *Yritysten Innovaatioympäristöt:*

Tutkimus Yritysten Innovaatiotoiminnasta ja Alueellisesta Innovaatiopolitiikasta Pirkanmaalla ja Keski-Suomessa (Helsinki: Tekes, Teknologiakatsauksia 120 / 2002).

Kolehmainen, J. (2004), 'Instituutioista yksilöihin: Paikallisen innovaatioympäristön kolme tasoa', in Sotarauta, M. and Kosonen, K-J. (Eds), Yksilö, *Kulttuuri ja Innovaatioympäristö: Avauksia Aluekehityksen Näkymättömään Dynamiikkaan* (Tampere: Tampere University Press).

Kolehmainen, J., Kautonen, M. and Koski, P. (2003), 'Korkeakoulut ja Alueellisen Innovaatiopolitiikan Visiot', in Aittola H. (ed.) *€KG? €urooppa, korkeakoulutus, globalizaatio?* (Jyväskylä: Jyväskylän yliopisto, Koulutuksen tutkimuslaitos).

Kuusisto, J. and Meyer, M. (2003), *Insights into Services and Innovation in the Knowledge Intensive Economy* (Helsinki: Tekes, Technology Reviews 134 / 2003).

Lechner, C. and Dowling, M. (2003), 'Firm Networks: External Relationships as Sources for the Growth and Competitiveness of Entrepreneurial Firms', *Entrepreneurship and Regional Development* 15, 1–26.

Leiponen, A. (2000), *Innovation in Services and Manufacturing: A Comparative Study of Finnish Industries* (Helsinki: ETLA, The Research Institute of the Finnish Economy, Taloustieto Oy).

Miettinen, R. (2002), *National Innovation System: Scientific Concept or Political Rhetoric* (Helsinki: Edita).

Miettinen, R., Lehenkari, J., Hasu, M. and Hyvönen, J. (1999), *Osaaminen ja Uuden Luominen Innovaatioverkoissa* (Helsinki: Sitra 226, Taloustieto Oy).

Muller, E. and Zenker, A. (2001), 'Business Services as Actors of Knowledge Transformation: The Role of KIBS in Regional and National Innovation Systems', *Research Policy* 30, 1501–1516.

Nachum, L. and Keeble, D. (2003), 'Neo-Marshallian Clusters and Global Networks: The Linkages of Media Firms in Central London', *Long Range Planning* 36, 458–480.

Nählinder, J. (2002), 'Innovation in KIBS: State of the Art and Conceptualizations', Prepared for the SIRP seminar, January 15. Available at the URL address: http://www.vinnova.se/innsysforsk/Slutrapporter%202002/KIBS%20f%F6rstudie. pdf. Accessed on 9 January 2004.

Piirainen, T. and Järvensivu, A. and Kautonen, M. (2000), *Asiantuntemus Kasvualana: Tuotteistaminen, Innovaatiotoiminta ja Kansainvälistyminen Pirkanmaan Osaamisintensiivisissä Yrityspalveluyrityksissä.* Unpublished final report. Tampere Technology Centre Ltd.

Porter, M. (1990), *The Competitive Advantage of Nations* (London: Macmillan).

Porter, M. (1998), 'Clusters and the New Economics of Competition', *Harvard Business Review* (November-December), 77–90.

Raunio, M. (2001), *Osaajat Valintojen Kentällä: Helsingin Tampereen, Turun, Jyväskylän, Porin ja Seinäjoen Seutujen Vetovoimaisuus Virtaavassa Maailmassa* (Tampere: Tampereen yliopisto, Alueellisen kehittämisen tutkimusyksikkö, Sente-julkaisuja 11/2001).

Rosenfeld, S.A. (1996), 'Does Cooperation Enhance Competitiveness – Assessing the Impacts of Interfirm Collaboration', *Research Policy* 25: 2, 247–263.

Schienstock, G. and Hämäläinen, T. (2001), *Transformation of the Finnish innovation system: A Network Approach* (Helsinki: Sitra, Sitra Report Series 7).

Storper, M. and Venables, A.J. (2002), 'Buzz: The Economic Force of the City', presented at the DRUID Summer Conference on "Industrial Dynamics of the New and Old Economy – Who is Embracing Whom?" Copenhagen/Elsinore, June 6–8.

Storper, M. and Venables, A.J. (2003), 'Buzz: Face-to-Face Contact and the Urban Economy', presented at the DRUID Summer Conference 2003 on "Creating, Sharing and Transferring Knowledge. The Role of Geography, Institutions and Organizations." Copenhagen, June 12–14.

Varis, J., Virolainen, V-M. and Puumalainen, K. (2004), 'In Search for Complementarities: Partnering of Technology-Intensive Small Firms', *International Journal of Production Economics* 90: 1, 117–125.

Chapter 14

Far Away, So Close? Regional Clustering of Mail Order Firms and Related Business Services in the Lille Metropolitan Area

Christian Schulz, H. Peter Dörrenbächer and Christine Liefooghe

With 175 firms and about 28,500 jobs in the mail order industry, the Nord-Pas-de-Calais region is by far the most important location in France for this sector, and even at the European level this concentration is unique. It represents about 65 per cent of this industry's activities in France (15 per cent of the European market). The two major groups in the French distance selling, *La Redoute (Redcats)* and *Les Trois Suisses*, as well as the main subsidiary of the German *Quelle* and many other mail order firms are located in the Lille Metropolitan Area (Figure 14.1).

At first glance, it might seem contradictory that an industry specialized in *distance* selling could benefit from spatial *proximity* on the supply side respectively from a specific institutional environment. While it is purported to be a footloose industry depending only on communication and transport infrastructure, obviously other location factors led to this cluster building. But it would be limiting to quote only historical backgrounds of this industry which succeeded the highly concentrated textile industry. This admittedly plays a crucial role and was undoubtedly the starting point for this development (see below), but the development of this cluster seems far more complex and influenced by a wide range of factors.

This final chapter adopts evolutionary economics to conceptualize the rise of this industry and to shed some light on current tendencies in this very dynamic industry. We have undertaken exploratory research based on conceptual literature, firm documents, former studies on the Lille Metropolitan Area and its mail order firms as well as interviews conducted in the region. After some general remarks on the use of evolutionary approaches in economic geography and a summary of its main aspects (Section 2), we present the study area and the rise of the mail order industry and its historical background (Section 3). Section 4 details particularities of the production system and its regional organization and exemplifies its specific environment, including current trends within the industry and their consequences for the cluster. The conclusions in Section 5 will detail the most promising research questions for our further work.

Figure 14.1   The Nord-Pas-de-Calais-Region and the Lille Metropolitan Area in France

Conceptual Framework: Evolutionary Perspectives in Economic Geography

Based on a fundamental critique of the static and a-historical character of neoclassic economic theory, evolutionary economics uses behavioural approaches and the concept of bounded rationality to explain historical development paths of specific firms, industries or regions – the latter being a main interest of economic geographers and increasingly taken into account by other scholars. The main characteristics of evolutionary theories can be summarized as follows (for more detailed explanations see Boschma and Lambooy 1999; Schamp 2002):

- economic development is generally *path dependent* and strongly influenced by individual/entrepreneurial decisions, technical and organizational innovations, and institutional settings;
- within a development path, different variations can occur and are subject to *selection processes*; from a more narrow evolutionist point of view, variation or diversity is even the crucial precondition for selective evolution (cf. Grabher and Stark 1997);
- a (irreversible) *lock-in* in a particular path (e.g. technology or production mode) stabilizes the selection result and is often related to *increasing returns* for other firms imitating (consciously or not) successful strategies (cumulative causation); on the other hand, lock-ins may turn out to have negative implications on the innovativeness of industries/firms/regions if they fall into the 'trap of rigid specialization' (cf. Grabher 1993);
- at the beginning of successful developments, *small events* play a crucial role; they are rather intentional actions than historic accidents or contingencies, but a certain notion of *randomness* or *chance* seems inherent to most cases (cf. Storper 1988).

As Hayter (2004, 98) pointed out, there is a narrow link between evolutionary perspectives and approaches in dissenting institutionalism (or 'old institutionalism', cf. Schamp 2002), since evolutionary explanations are 'also embedded in social and political as well as economic processes'. Thus 'institutional trajectories' are 'inevitably path dependent' themselves (ibid), and '[b]luntly stated, new incremental, major and radical 'hard' technologies require new incremental, major and radical institutional arrangements, habits, routines, values and conventions' (ibid,103). Hence, besides the pure economic development path related to innovations and organizational changes and their regional implications, institutions such as collective conventions, untraded interdependencies, increasing returns (see above) and further aspects of embeddedness have to be considered as path dependent. In the same way, aspects of collective learning, interorganizational transfer of (tacit) knowledge and their relation to spatial and other forms of proximity have to be taken into account (Schamp 2002, 47 et seq.). On the other hand, institutions are themselves path determining with regard to firm behaviour and the development of industrial sectors, which leads Hayter (2004, 104) to refer to Myrdal's concept of circular interdependence and cumulative causation.

In recent years, evolutionary approaches have become more popular in economic geography since they provide interesting models and explanations for the analysis of and inquiry into the emergence of spatially concentrated industries or other spatial differentiations within the economy. However, standard models in evolutionary economics (cf. Nelson and Winter 1982, Dosi et al. 1988, Cimoli and Dosi 1995) focused on technological change and innovativeness within certain industries, without really considering how geographic relationships influence firms' competitiveness and learning capacities. Recent works in economic geography (e.g. Grabher and Stark 1997; Boschma and Lambooy 1999; Schamp 2000; 2002; Hayter 2004) discuss the obvious and fruitful linkages between evolutionary perspectives and current approaches in our discipline like localized production systems, embeddedness, regional learning processes and so on.

While the conceptual debate on evolutionary approaches in economic geography has become more dynamic in recent years, there is still a lack of application and empirical studies using this perspective. Most of the empirical work done so far dealing with path dependent spatial developments (e.g. Boschma 2003; Moßig 2000; Klepper 2002) has studied localization patterns of manufacturing industries, granting service industries only minor roles. One of the few exceptions[1] is Boschma's and Weltevreden's (2004) study on the evolutionary nature of e-commerce in inner cities. We will refer to this study later in our analysis when dealing with recent developments in the mail order industry which is strongly interwoven into internet retailing. But first we present the specific situation in our study area.

Regional Pathways in the Lille Metropolitan Area: From a Textile Industry Region to the 'Mail Order Valley Europe'

> *... history becomes the raw material for new dynamics.*
>
> (Boschma and Lambooy 1999: 415)

The unique concentration of mail order firms in Northern France and especially in the Lille Metropolitan Area results from a profound restructuring process that followed the deep crisis of traditional industries in this region, of the textile industry in particular. We will illustrate the current situation in a more detailed manner before we elaborate on its historical background and the different trajectories that emerged from the regional economic decline.

Current Situation

As Table 14.1 shows, eight out of the fifteen most important mail order firms in France are located in the Nord-Pas-de-Calais region. Among them, two companies are outstanding, not only at the national scale: *Les Trois Suisses* and *La Redoute* (*Redcats*). The majority shareholder of *Les Trois Suisses* is the German company *Otto Versand*, the largest in the world (followed by Quelle). *La Redoute* was ranked fourth worldwide in 1998 and has become third due to a variety of acquisitions mainly in North America (Morganosky and Fernie 1999, 275; Redcats 2004). This growth has been accompanied by a renaming in *Redcats*, combining the two syllables 'Red' from *La Redoute* and 'Cats' from catalogue retail, in order to 'encapsulate the company's new corporate identity and its growing international stature' (Redcats 2004a).

---

1   See also Bathelt (2001) on Leipzig's media industry cluster.

Table 14.1 Ranking of France's 15 most important mail order firms in 2000

Rank	Company	Affiliation	Activity	Annual Turnover (€m, Mail Order only)
1	**La Redoute**	Redcats	General Assortment	1,452
2	**Trois Suisses France**	Trois Suisses International	General Assortment	1,182
3	CAMIF	CAMIF	General Assortment	690
4	**Quelle La Source**	Quelle France	General Assortment	398
5	**Movitex**	Redcats	Textile	326
6	**Blanche Porte**	Trois Suisses International	Textile	237
7	**La Maison de Valérie**	Redcats	Housewares	207
8	Yves Rocher	Yves Rocher	Cosmetics	199
9	**Damart Serviposte**	Damart	Textile	153
10	Dial		Music, Videos	150
11	**Sadas Vert Baudet**	Redcats	Textile	142
12	France Loisirs	France Loisirs	Books, Music, Videos	128
12	Sélection du Reader's Digest	Reader's Digest	Books, Music, Videos	128
13	Editions Atlas		Books, Music, Videos	126
14	Europe Epargne		General Assortment	118

Boldface denotes headquarters located in Nord-Pas-de-Calais.

Source: ARDNPC 2001: 3.

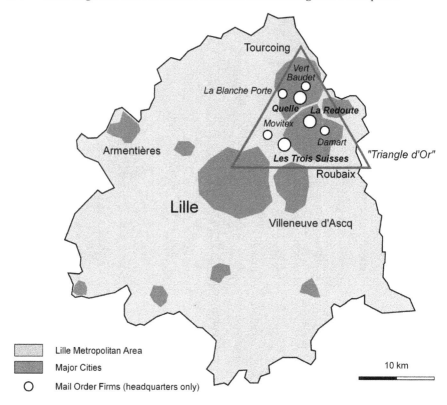

Figure 14.2   Localization of important mail order firms in the study area

More than 95 per cent of the regional employment in the distance selling industry is concentrated in the two cities of Roubaix and Tourcoing (Guilbeau 1992, 5). This eastern part of the Lille Metropolitan Area is also called 'the golden triangle of distance selling' ('Triangle d'Or', 34). With the activities of the major firms and their functional reorganization due to new and lean management strategies, a wide variety of narrowly linked service firms and specific suppliers have emerged in this region. In addition to its traditional linkages to the transportation and logistics sector, the mail order industry is today strongly dependent on highly specialized service providers such as product photographers, design agencies, printing shops (for catalogues etc.), and financial services (for leasing contracts). Furthermore, ICT-services have become more important due to an obvious convergence between traditional forms of mail order and new types of e-commerce.

Historical Background

The profound and ongoing restructuring process of the regional economy started much earlier than the actual decline of the formerly dominant textile industry.

Even in the late 19th century, some textile companies diversified their activities towards related (e.g. garment retailing, mechanical engineering) or less related (e.g. publishing companies, newspapers) sectors. But generally speaking, the textile industry has known a positive and relatively stable development for more than a century – even though the change from wool and linen to cotton as the predominant raw material has caused substantial change. This period has influenced the industry as well as the region, since the latter is characterized by the rising of a limited number of successful entrepreneurial families which soon dominated the market. Only a few entrepreneurs from other regions succeeded in penetrating their very close and tied interfirm patterns. On the contrary, the well established entrepreneurs pursued a marriage policy that further tightened the links between the regional firms and strengthened their community (see below for the still considerable importance of some of these family owned companies in the French economy; further details are exhaustively presented in Pouchain 1998).

World War I, however, heavily affected the border region of Nord-Pas-de-Calais, not only in its general economic impacts, but particularly in the confiscation of industrial equipment during the German occupation as well as the bomb damage to the residential and industrial fabric. Therefore it is not very surprising that soon after 1918, local firms began to explore other fields of activity. This was the time when *Phildar* was founded, the predecessor of the trade chain *Auchan*, nowadays one of the big players on the European retail market. The real estate sector became an important activity as local firms began to reuse and deal with ancient industrial land and buildings. The mail order industry also has its origins in the turbulent 1920s and 1930s (see below). After World War II, the retailing and the mail order companies expanded and were imitated by many other newly established firms. In addition, other new businesses like the hotel industry (e.g. the *Accor* group with its chains *Mercure, Novotel, Ibis, Formule 1* etc.) and the automotive supply industry grew impressively (Dörrenbächer and Schulz 2005).

The Rise of the Mail Order Industry

The factors for the rise of this industry can most easily be exemplified by the trajectories of the two earliest and still most important French companies, *La Redoute / Redcats* and *Les Trois Suisses* (see Table 14.2).

La Redoute.   *Les Filatures de la Redoute*, founded in 1873 by the Pollet family, had a long tradition in the Roubaix wool industry (spinning and weaving mills). It was literally by chance that in 1922 the Pollets entered the mail order market. After an unsuccessful deal with a garment-producing client firm, they had to reduce their redundant stock of wool, and – necessity being the mother of invention – decided to turn towards retailing. A first attempt at selling knitting wool directly to individual (female) consumers was very successful, placing a small advertisement (4 cm²!) in

Table 14.2   Key facts of the two most important mail order firms in France

	Redcats La Redoute	Les Trois Suisses International
Annual turnover (million Euro)	4,400*	2,600**
Annual turnover in France	2,008*	1,130**
Number of employees worldwide	19,000	10,700
Client contacts per year	150 million	70 million
Number catalogues per year	22	300
Number of parcels sent per year	85 million	60 million
Share of textile products	64%	70%
Selling channels:		
- telephone	82.5%	52.4%
- mail order		34.4%
- internet/minitel	14.1%	10.6%
- local shops	3.4%	2.6%
Number of commercial web sites	51	35

*2003, **2001

Sources: ARDNPC 2001; Les Trois Suisses 2004b; Redcats 2004b.

a local newspaper (*Journal de Roubaix*), thereby eliminating wholesalers or other intermediaries involved in the trade. Three years later, the firm edited a monthly journal (*Penelope*) presenting fashion products. When the first catalogue (16 pages, textiles only) was published in 1928, *La Redoute* already had about 600,000 clients. This success story continued after Second World War, and led the firm to even stop its own production in 1960 and to concentrate on its mail order activities only. In the early 1960s, *La Redoute* was one of the early adapters of new computer aided techniques and robots for their inventory management and packaging which considerably reduced the time of delivery.[2] In 1970 the traditional mail order business was extended by a multichannel selling strategy (see below), i.e. by the implementation of a first call centre for telephone orders and by the opening of more than 100 agencies/chain stores all over France. In the 1980s and early 1990s, *La Redoute* took over a series of important French mail order firms like *Movitex (*1983), *Le Vert Baudet* (1989), and *La Maison de Valérie* (1991) (Table 14-1). In 1994, *La Redoute* merged with the Franco-Swiss *Pinault-Printemps* group. The increased investment capital stock enabled the group to realise numerous further acquisitions since 1995, mainly in North-America, Great Britain and Scandinavia. Today, more than 50 per cent of its sales are conducted outside France (Redcats 2004a; Guilbeau 1992, Pouchain 1998).

Les Trois Suisses. At the beginning of the 1930s, well-established and large textile firms in Roubaix and Tourcoing ran into difficulties due to decreasing demand and general economic uncertainty. Thus Xavier Toulemonde, inheritor of the Toulemonde-Destombe spinning company, decided in 1932 to imitate the Pollet brothers and also started – with four employees – a mail order service for knitting wool which was called *Les Trois Suisses*. Legend has it that the only packer carried the parcels by bike to the local mail office. The ensuing success was comparable to the *Redoute* story, and the number of employees increased rapidly to 100 in 1936 and to 238 in 1938. In 1934, the first foreign subsidiary was founded in Belgium. Since 1974, *Les Trois Suisses* have cooperated with the German company Otto-Versand, now majority shareholder (50 per cent) of the company.[3] In 1968, the company's first call centre was opened, and from 1983 ordering via the French *Minitel* system was possible (Les Trois Suisses 2004a; Guilbeau 1992; Pouchain 1998). The firm expanded continously through the acquisition of other French mail order firms (*La Blanche Porte* 1983; *Becquet* 1987) and by founding various subsidiaries in European and overseas countries. However, it was only in 1991 that the company

---

2   From 1984 on, *La Redoute* was able to offer a 48 hour dispatch service and to guarantee a full refund in case of delay; in 1995, the maximum delivery time was reduced to 24 hours after ordering.

3   The second largest shareholder is the local Mulliez family (45%), owner of the French *Auchan* and other important trade chains; the '*Auchan* empire' is also a result of successful diversification strategies in the textile industry (cf. Pouchain 1999, 344–347).

closed the last spinning plant and thus abandoned its still working textile industry branch.

Specific Aspects of the Production System

Before we discuss the mail order value chain in more detail, the industry has to be defined more precisely, in particular with regard to the often synonymously used term 'distance selling'. The latter is the generic term for different kinds of trade dealing with remote clients. According to the European Commission, a *distance contract*:

> means any contract concerning goods and services concluded between a supplier and a consumer under an organized distance sales or service provision scheme run by the supplier, who, for the purpose of the contract, makes exclusive use of one or more means of distance communication up to and including the moment at which the contract is concluded.
>
> (European Commission 1997, 4)

Hence, besides the traditional mail order business, distance selling comprises also internet commerce and other forms of direct marketing (e.g. personal direct mailing, telephone marketing), and even door-to-door distribution (i.e. the sales agent is considered as being a 'means of distance communication' used by the supplier). Generally speaking, no direct face-to-face contact between the supplier and the consumer takes place. Due to the convergence of traditional mail order services and more recent forms of e-commerce etc., it seems appropriate to apply the more general term of distance selling.[4] In this study, however, we prefer a distinction and will mainly use the term mail order meaning the traditional type of distance selling.

The Value Chain of Mail Order Services

The value chain of the Nord-Pas-de-Calais mail order industry is characterized by numerous backward and forward linkages within and beyond the region. On the supply side, both external suppliers as well as company-owned producers deliver the goods and services offered in their catalogues. With the exception of some separate or decentralized distribution centres (e.g. *Quelle* in Orléans), all goods are delivered to the companies' warehouses where they are stocked and prepared for further processing (Figure 14.3).

Some of the mail order firms still own production units in the garment industry, but most of the goods come from external suppliers. The clients' orders reach the provider via different channels (see below), and are processed and lead to the

---

4   With the same thrust, the French term 'vente par correspondance' (VPC) is increasingly replaced by 'vente à distance' (VAD); sometimes an abbreviated combination of both is used: VPC/VAD

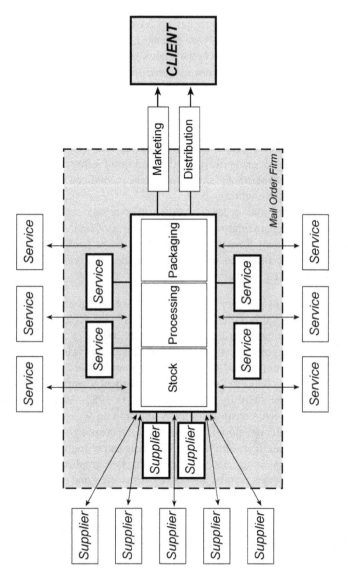

Figure 14.3   The value chain of the mail order industry

packaging before the goods are delivered to the client. Within this process, as well as in the marketing realm, more and more tasks have been externalized to specific business services. In addition, general (public) services such as the national railway company *SNCF* and the French postal service (*La Poste*) are used and have adapted to the specific needs of the mail order firms The *SNCF*, for example, through its transportation subsidiary *SERNAM* operates no fewer than three freight stations in the Roubaix-Tourcoing area, and *La Poste* manages two distribution centres for the millions of parcels and catalogues sent to the clients each year. It should be mentioned that more than 10 per cent of all parcels distributed in France are dispatched by the mail order firms in the Lille Metropolitan Area (Guilbeau 1992).

From an organizational point of view, four types of service firms can be distinguished:

- independent providers offering specialized services (e.g. design, printing shops), in general not exclusively for the mail order industry;
- formerly internal departments that have been externalized (e.g. management buy-outs and subcontracting);
- firms created by the mail order companies offering services outside their core business, but complementary to it (e.g. financial services);
- formerly independent firms incorporated into the respective group through mergers and acquisitions in order to control their activities and to reduce transaction costs (e.g. logistics).

Most of these service providers are located close to the headquarters and formed a dense supplier network in this region. About 900 firms in the region are supposed to work at least partly for the mail order industry (FACE 2003, 27).

Among the most important company owned service providers, the following examples should be mentioned. In 1970, *La Redoute* founded the *Finaref/Kangourou* financial service[5] offering loans to its clients to enable them to purchase its products. *Les Trois Suisses* did the same by creating *Cofidis* in 1982. As for logistics, *La Redoute* incorporated a local transport company and built up *Sogep*. In the marketing realm, the most spectacular development certainly is the creation of the design centre *La Cité Numérique* by *Les Trois Suisses* in 1996. This centre regroups several firms and departments such as design agencies, product photographers (50,000 photos taken per year), multimedia experts, marketing firms etc., and was conceived as a central service unit (one stop shopping) for all mail order firms belonging to the *Trois Suisses International* group (LCN 2004). Two hundred fifty employees are working for *La Cité Numérique* which, in the meanwhile, has established a second site at Paris[6] and is now also working for clients outside the mail order industry (e.g. *Air France, La Poste, Renault*). The same strategy was followed when *Les Trois Suisses* founded *Mondial Relay* as the group's exclusive packaging service.

---

5   Today belonging to the Crédit Agricole Group
6   Through the acquisition of the firm 'Web Valley' in 2001, strengthening its expertise in the internet business.

The Local Institutional Environment

The aforementioned network of various service firms is accompanied by several (mainly public) initiatives to improve the local environment for the mail order industry. Three areas of activities should be exemplified: strategies for regional marketing and specific support for SME, professional and higher education, and the organization of professional gatherings.

Regional marketing and support for SMEs. Under the 'catchy' label *Mail Order Valley Europe* (MOVE), a wide range of private and public actors in the region have created a network whose major objectives are:

- To contribute to the development of multi-channel Distance Selling;
- To act as a link between the different actors of the Distance Selling industry;
- To aid in the transfer of knowledge and information.

(MOVE 2004)

The network comprises local and regional public authorities, the local chambers of commerce and other business associations, several institutions of higher education, *La Poste* as well as the *Lille Grand Palais* convention center. It is mainly funded by the participating municipalities and the two Départements of the Nord-Pas-de-Calais region. Its *MOVE CENTER* offers training courses particularly for SMEs to enable them to adapt to technical innovations and changes in the mail order market. Its newsletter tries to foster communication within the industry. Furthermore, it is one of the major organizers of the annual distance selling congress held in Lille since 1997, *Les Rendez-Vous de la Vente à Distance et du Marketing Direct*.

Professional and higher education. Several institutions of professional and higher education are offering specific study programs and courses in the realm of distance selling and direct marketing. Undoubtedly the most prominent one is the *MD-Lab* (Direct Marketing Laboratory), established by the renowned *Lille Graduate School of Management (École Supérieure de Commerce de Lille, ESC[7])* and co-funded by the *European Regional Development Fund (ERDF)*. It offers an MBA program in direct marketing and e-commerce and is supposed to be the most reputable European school in this realm. It narrowly co-operates with the firms located in the region both in terms of recruiting external teaching staff as well as in terms of internships offered to the graduate students.

Professional gatherings. The annually-held European Distance Selling and Direct Marketing Rendez-Vous in Lille is organized by MOVE, FEVAD (the French association of distance selling) and AEVPC/EMOTA (the European association).

---

7   The ESC in France are privately conducted business schools which are part of the French 'Grandes Ecoles' system of elite schools.

The main sponsor is – unsurprisingly given the importance of its mail order clients – La Poste. For more than eight years, this conference, held in the main convention centre of the region, Lille Grand Palais, has gathered 200 exhibitors and 8,000 visitors for three days. According to Maskell et al. (2004) it therefore forms a temporary cluster of experts in this industry. Its importance for the knowledge exchange and the innovativeness of this industry can be considered as an important supplement to the permanent cluster of firms in the study area, or as Maskell et al. put it, they are 'a bit like close cousins' (ibid, 24). In addition, the fact that Lille is the uncontested location in France to host this international exhibition underlines the importance of this cluster.

Current Trends

As indicated above, there is a certain convergence between the traditional mail order business and new forms of internet commerce. Yet, 'catalogue firms have been most successful to embrace e-commerce' (Boschma and Weltevreden 2004, 6). Among the ten most important internet commerce firms in France, three mail order firms are listed, together with the main actors of the national tourism and media/music industry (ARDNPC 2001, 13). Due to the new information and communication technologies, the mail order firms have not only become able to use further channels for marketing, product presentation and selling, but also to improve their processing and transport logistics. Furthermore, the new technologies also allow new forms of contacts between supplier and client. So-called consumer response strategies in the retail sector have primarily been developed by the mail order industry and have lead to an *ultra personalization* of the client relationships. Data on the individual profile of the consumers are systematically analysed and used for adapted marketing strategies (cf. EAN 2004; ECR 2004).

Conclusions and Research Agenda

The rise of this unique mail order industry cluster and its continuous adaptation to new market requirements, technological changes and other factors offer a promising playground for the empirical application of evolutionary approaches. As we could see in this rather exploratory study, a mix of historic incidents and proactive strategies, both strongly embedded in the local economic and social environment, has marked the development of this industry. However, the first results presented in this paper will have to be verified by a more profound and diversified analysis of this evolution. According to Boschma and Lambooy (2004, 422), the *trigger* for the respective innovations and adaptations can be far more complex than discussed here and merits a much closer look at the *small events* and their impacts on the regional economy.

Consequently, the following questions should guide our further research in this case study:

- Who are the key actors during the rise of the local mail order industry? When did they take which kind of crucial decision? Which other options did they pursue and abandoned, for which reasons?
- What kind of structural persistence (financial, organizational, social ...) can be identified over the restructuring process?
- How do the suppliers and service providers contribute to the quality of the location, and how do they profit from the focal mail order firms?
- How do they interact and which types of knowledge transfer and collective learning can be observed?
- Why did other mail order firms and related service providers choose the region when (re)localising?
- What is the role of public authorities and other intermediary organizations/ institutions? How important are local research and higher education institutions?
- How are the local entrepreneurs interwoven into the regional and national policy?
- How does the mail order cluster adapt to new challenges and ongoing concentration and internationalization processes within the industry?
- To what extent do new information and communication technologies foster a spatial reorganization of the mail order industry, i.e. do locational / proximity advantages become less important?
- What kind of changes can be observed regarding the specific value chain of this industry?
- Does the sector interact or even converge with other important industries in the region (e.g. the big trade chains) and what kind of synergies might be used?
- What are the general perspectives for the regional economy?

The main challenge will certainly be to operationalize these research questions regarding both an appropriate methodology (certainly with a qualitative thrust) and a sustainable access to the main actors related to the cluster. It is also planned to take into account the evolution in the German mail order industry since it is strongly connected to the French market. A strictly comparative study, however, seems inadequate given the uniqueness of the situation in Northern France.

For our further research in this field, the recent trends and the foreseeable changes in this industry are of a particular interest. Thus we can examine the rise of this industry not only from an historical perspective, but also to analyse the current adaptation strategies of the mail order business in real-time. This will probably allow us to identify misleading paths and strategic failures, often more difficult to detect in retrospective studies.

Acknowledgements

The authors would like to thank Dr. Andrew Murphy, University of Birmingham, for his most appreciated help while revising and spell-checking a previous version of this paper.

References

ARDNPC (2001), Agence Régionale de Développement du Nord-Pas-de-Calais: Vente à distance, de la VPC au commerce électronique. *La Note d'information Economique* 231. http://www.ardnpc.org/noteco/actu/frames_note.asp?case=231 (retrieved 2004/07/31)

Bathelt, H. (2001), *The Rise of a New Cultural Products Industry Cluster in Germany: The Case of the Leipzig Media Industry*. IWSG Working Papers 6–2001. Frankfurt a.M. (Johann Wolfgang Goethe-Universität, Institut für Wirtschafts- und Sozialgeographie)

Boschma, R.A. and Lambooy, J.G. (2004), 'Evolutionary Economics and Economic Geography', *Journal of Evolutionary Economics* 9, 411–429.

Boschma, R.A. and Weltevreden, J.W.J. (2004), 'The Evolutionary Nature of B2C E-Commerce in Inner Cities', in Taylor, M. (ed.), *E-commerce, E-business and the Dynamics of Economic Development* (Aldershot: Ashgate).

Cimoli, M. and Dosi G. (1995), 'Technological Paradigms, Patterns of Learning and Development: An Introductory Roadmap', *Journal of Evolutionary Economics* 5, 243–268.

Dörrenbächer, H.P. and Schulz, C. (2005), 'Dienstleistungsstandort Nord-Pas-de-Calais – Hoffnungsschimmer im Strukturwandel einer Altindustrieregion', *Geographische Rundschau* 57: 9, 12–18.

Dosi, G., Freeman, C., Nelson, R., Silverberg, G. and Soete, L. (eds) (1988), *Technical Change and Economic Theory* (London, Pinter).

EAN (2004), European Article Numbering International. http://www.ean-int.org/ (retrieved 2004/07/31).

ECR (2004), Efficient Consumer Response. http://www.ecr.de/ (retrieved 2004/07/31).

European Commission (1997), *Directive 97/7/EC of the European Parliament and of the Council of 20 May 1997 on the Protection of Consumers in Respect of Distance Contracts* (Brussels: European Commission).

FACE (2003), 'FACE : Le Magazine des Entreprises de Lille Métropole 158', *Marketing/VAD: Développer une Relation Privilégiée avec le Client* (Lille).

FEVAD (2004), 'Fédération des Entreprises de Vente à Distance', *Chiffres clés 2004 – Vente à Distance – E-Commerce* (Paris).

Freel, M.S. (1998), 'Evolution, innovation and learning: evidence from case studies', *Entrepreneurship and Regional Development* 10: 2, 137–149.

Grabher, G. (ed.) (1993), *The Embedded Firm: On the Socioeconomics of Industrial Networks* (London: Routledge).

Grabher, G. and Stark, D. (1997), 'Organizing Diversity: Evolutionary Theory, Network Analysis and Postsocialism', *Regional Studies* 31: 5, 533–544.

Guilbeau, N. (1992), *La Vente par Correspondance à Roubaix – Tourcoing : Nature, Structure et Enjeux*. Mémoire de DEA Université des Sciences et Technologies de Lille (unpublished).

Hayter, R. (2004), 'Economic Geography as Dissenting Institutionalism: The Embeddedness, Evolution and Differentiation of Regions', *Geografiska Annaler* 86B: 2, 95–115.

Klepper, S. (2002), 'The Capabilities of New Firms and the Evolution of the US Automobile Industry', *Industrial and Corporate Change* 11: 4, 645–666.

LCN (2004), *La Cité Numérique – Déclencheur d'Actes d'Achats: Dossier de Presse* (Croix: LCN).

Les Trois Suisses (2004a), 'Les Origines des 3 Suisses', http://www.3suisses.fr/emplois/3S/histoire.htm (retrieved 2004/07/31).

Les Trois Suisses (2004b), 'Quelques Chiffres Clés', http://www.3suisses.fr/emplois/3S/chiffres.htm (retrieved 2004/07/31).

Maskell, P., Bathelt, H. and Malmberg, A. (2004), 'Temporary Clusters and Knowledge Creation: The Effects of International Trade Fairs, Conventions and Other Professional Gatherings', *SPACES* 2004-4 (Marburg: Universität Marburg, Geographische Fakultät).

Morganosky, M.A. and Fernie, J. (1999), 'Mail Order Direct Marketing in the United States and the United Kingdom: Responses to Changing Market Conditions', *Journal of Business Research* 45, 275–279.

Moßig, I. (2000), 'Lokale Spin-off-Gründungen als Ursache räumlicher Branchencluster. Das Beispiel der deutschen Verpackungsmaschinenbau-Industrie', *Geographische Zeitschrift* 88: 3–4, 220–233.

MOVE (2004), 'Mail Order Valley Europe: MOVE / MOVE CENTER', Presentation sheet (Lille: MOVE).

Nelson, R.R. and Winter S.G. (1982), *An Evolutionary Theory of Economic Change* (Cambridge, Mass.: Belknap-Harvard Univ. Press).

Pouchain, P. (1998), *Les maîtres du Nord – du XIXe siècle à nos jours* (Paris: Perrin).

Redcats (2004a), 'Historique du Groupe', http://www.redcats.com/fr/groupe/chiffre.htm (retrieved 2004/07/31).

Redcats (2004b), 'Chiffres Clés', http://www.redcats.com/fr/groupe/histo.htm (retrieved 2004/07/31).

Schamp, E.W. (2000), *Vernetzte Produktion. Industriegeographie aus institutioneller Perspektive* (Darmstadt: Wissenschaftliche Buchgesellschaft).

Schamp, E.W. (2002), 'Evolution und Institution als Grundlagen einer dynamischen Wirtschaftsgeographie: Die Bedeutung von externen Skalenerträgen für geographische Konzentration', *Geographische Zeitschrift* 90: 1, 40–51.

Storper, M. (1988), 'Big Structures, Small Events, and Large Processes in Economic Geography', *Environment and Planning A* 20, 165–185.

Index

For Product Safety Concerns and Information please contact our EU
representative GPSR@taylorandfrancis.com
Taylor & Francis Verlag GmbH, Kaufingerstraße 24, 80331 München, Germany

www.ingramcontent.com/pod-product-compliance
Lightning Source LLC
Chambersburg PA
CBHW071405050326
40689CB00010B/1756

* 9 7 8 1 1 3 8 2 7 5 5 7 7 *